PHYSICAL GEOLOGY AND TECTONICS

Dr. Laxman Siyol & Dr. B. R. Rojh

BlueRose ONE.com
Stories Matter
NewDelhi • London

BLUEROSE PUBLISHERS
India | U.K.

BlueRose ONE.com
Stories Matter
NewDelhi • London

For permissions requests or inquiries regarding this publication,
please contact:

BLUEROSE PUBLISHERS
www.BlueRoseONE.com
info@bluerosepublishers.com
+91 8882 898 898
+4407342408967

ISBN:978-93-6783-749-8

Cover design: Shubham
Typesetting: Sagar

First Edition: December 2024

Physical Geology and Tectonics

The book provides a comprehensive understanding of geology and planetary sciences, focusing on Earth's physical structure and the processes shaping our planet. It begins with an introduction to geology, detailing its branches, scope, and applications in various fields. The book then transitions to planetary sciences, discussing the origins of the universe and the solar system, meteorites, and impact craters. It covers Earth's unique characteristics within the solar system, including its size, shape, mass, and gravitational parameters. The book also discusses seismology, the internal structure of Earth, earthquakes, and geological processes on a grand scale, such as continental drift, sea-floor spreading, and plate tectonics. It also examines the origin of oceans, continents, and mountains, and the effects of weathering, erosion, wind, rivers, glaciers, and underground water on landforms. The final chapters focus on Indian geomorphology, offering a regional perspective on global geological processes. This book is an essential resource for students, researchers, and professionals in geology and planetary sciences, combining theoretical knowledge with real-world examples to enhance understanding of both Earth and the wider solar system.

First Author, Dr. Laxman Siyol is an Assistant Professor (Guest Faculty) in Department of Geology, Jai Narain Vyas University, Jodhpur, Rajasthan (India). He completed his Ph.D. in Geology (Specialization in Economic mineralization and Petrology). During his three years of employment as a hydrogeologist at Water and Sanitation Management Organization, Gujarat, Dr. Siyol made substantial contributions to the management of groundwater resources. Apart from his teaching position, he owns Bharat Bhujal Survey, a government-approved business that specializes in groundwater surveys. His background and skill set demonstrate a strong dedication to geological research and useful applications.

Second Author, Dr. B. R. Rojh is an Assistant Professor (Guest Faculty) in Department of Geology, Jai Narain Vyas University, Jodhpur, Rajasthan (India). He completed his M.Sc. in Geology, from Department of Geology, Government Dungar College, Bikaner and Ph.D. in Geology (Specialization in Hydrogeology) from Department of Geology, Jai Narain Vyas University, Jodhpur, Rajasthan (India). His education and experience show that he is very devoted to geological research and execution.

Preface

The study of geology forms the cornerstone of understanding Earth's physical structure and the dynamic processes that have shaped its evolution over billions of years. *Physical Geology and Tectonics* is a comprehensive guide designed to explore these concepts, seamlessly bridging foundational knowledge and advanced insights. The book introduces readers to the vast scope of geology, its branches, and its applications across diverse fields. From the origins of the universe and the solar system to Earth's unique position within it, this book delves into topics such as meteorites, impact craters, seismology, and the internal structure of Earth. It provides an in-depth analysis of geological phenomena, including continental drift, sea-floor spreading, plate tectonics, and the formation of oceans, continents, and mountains. In addition to these global geological concepts, the book focuses on the geological forces that shape the Earth's surface. Topics such as weathering, erosion, and the role of geomorphic agents like rivers, glaciers, wind, and underground water are explored in detail. The regional perspective provided through the study of Indian geomorphology enriches the reader's understanding of how global geological processes manifest in specific terrains. This emphasis on both global and regional geology ensures that the book is not only a theoretical resource but also a practical guide for understanding Earth's dynamic systems. The authors bring significant expertise and experience to this endeavor. Dr. Laxman Siyol, an Assistant Professor (Guest Faculty) in the Department of Geology at Jai Narain Vyas University, Jodhpur, holds a Ph.D. in Geology with a specialization in economic mineralization and petrology. He combines academic rigor with practical experience, having worked as a hydro geologist and managing Bharat Bhujal Survey, a government-approved company specializing in groundwater surveys. Dr. B.R. Rojh, also an Assistant Professor (Guest Faculty) in the Department of Geology at Jai Narain Vyas University, brings a wealth of academic and professional expertise. With a M.Sc. in Geology and a Ph.D. specializing in hydrogeology, his work reflects a deep commitment to advancing geological research and its real-world applications. This book is structured to provide a progressive learning experience, beginning with foundational principles and advancing to complex geological concepts. The combination of theoretical knowledge and practical examples makes this book an invaluable resource for students, researchers, and professionals. By offering insights into both global and Indian geological systems, *Physical Geology and Tectonics* aims to inspire curiosity, enhance understanding, and support the academic and professional growth of its readers.

Dr. Laxman Siyol
Dr. B. R. Rojh

Acknowledgement

The completion of this book, Physical Geology and Tectonics, represents a significant milestone in our academic journey. It would not have been possible without the unwavering support, guidance, and encouragement of several individuals, to whom we are deeply indebted. First and foremost, we have an opportunity to express our gratitude, but no words are adequate to convey the profound privilege and deep sentiments from the core of our hearts to our esteemed, venerated mentor, erudite and eminent scientist, **Professor (Dr.) K.L. Shrivastava,** Vice Chancellor of Jai Narain Vyas University, Jodhpur, India. His invaluable guidance, mentorship, and blessings have been a constant source of inspiration, helping us achieve every elevation in our academic pursuits. We extend our heartfelt thanks to **Professor S.R. Jakhar,** Head of the Department of Geology, Jai Narain Vyas University, Jodhpur, India. His insightful suggestions, support, and mentorship have been instrumental in shaping our understanding of geology and inspiring us to explore the dynamic processes that govern the Earth. We are deeply grateful to **Professor (Dr.) S.C. Mathur**, Former Head of the Department of Geology, Jai Narain Vyas University, Jodhpur, India, for his pioneering contributions to geology and his unwavering encouragement. His legacy as a scholar and mentor continues to inspire us and countless others in the field of geological sciences. Our deepest gratitude also goes to **Assistant Professor (Dr.) V. S. Parihar,** Department of Geology, Jai Narain Vyas University, Jodhpur, India, for his constructive feedback, meticulous guidance, and constant encouragement throughout the development of this book. His contributions have significantly enhanced the quality of this work. We also take this opportunity to express our sincere appreciation to our students, families, friends, and colleagues for their unwavering support and patience during the creation of this book. Their belief in our work has been a source of immense strength and motivation. This book is a culmination of years of dedication, academic rigor, and practical experience. We hope it serves as a valuable resource for students, researchers, and professionals, fostering a deeper understanding of geological principles and processes. We dedicate this book to all those who share a passion for unraveling Earth's mysteries and to those who inspire curiosity and academic growth through their relentless pursuit of knowledge.

Dr. Laxman Siyol
Dr. B.R. Rojh

Contents

Chapter 1
Geology and its branches, scopes, and applications

Geology is the scientific study of the Earth's composition, internal processes, and historical and contemporary changes. The purpose is to understand the Earth's composition, structure, and dynamic processes that have influenced the planet's interior and exterior. Geology encompasses numerous fields, each focusing on certain aspects of Earth science:

1. **Physical Geology:** A branch of geology focused on the examination of Earth's components and the processes affecting them. This encompasses the examination of rocks, minerals, and the processes behind geological events such as earthquakes, volcanoes, and erosion. Principal Subjects: Lithogenesis, tectonic plates, weathering processes, and erosion.
2. **Mineralogy:** A branch of geology focused on the examination of minerals, including their structure, characteristics, classification, and formation processes. Understanding the composition and behavior of rocks is essential. Principal Subjects: Mineral characteristics, crystallography, and mineral classification.
3. **Hydrogeology:** A branch of geology focused on the distribution and flow of groundwater. It examines factors such as aquifer characteristics, water quality, and

the influence of human activities on groundwater systems. Principal Subjects: Aquifers, underground water movement, and water quality.

4. **Historical Geology:** A branch of geology focused on comprehending the Earth's past via the examination of stratigraphy and paleontological remains. It aims to recreate historical habitats, climates, and biological evolution. Principal Subjects: stratigraphy, paleontology, and the geologic time scale.
5. **Paleontology:** A branch of geology focused on the examination of fossils and prehistoric organisms. It offers insights on historical ecosystems, evolutionary processes, and the chronology of life on Earth. Principal Subjects: Fossil identification, paleoenvironments, and extinction events.
6. **Structural geology:** A branch of geology that examines the deformation of rocks and the mechanisms that induce these alterations. It addresses characteristics like faults, folds, and strata, facilitating comprehension of the Earth's tectonic processes. Principal Subjects: Tectonic plates, faulting, folding, and orogenesis.
7. **Petrology:** A branch of geology that examines the origin, composition, and structure of rocks. Petrology is categorized into igneous, sedimentary, and metamorphic petrology, each focusing on distinct rock types and their genesis processes. Principal Subjects: Types of rocks, processes of rock formation, and categorization of rocks.
8. **Economic Geology**: A branch of geology that examines the origin, distribution, and use of Earth's mineral and energy resources. Principal Subjects: Ore Deposits, Mineral Exploration, and Resource Management.
9. **Sedimentology:** A branch of geology that examines the mechanisms of sediment generation, transport, and deposition. It assists in comprehending sedimentary rock strata and the historical events encapsulated within them. Principal Subjects: Sediment Classification, Transport Mechanisms, Deposition Processes, Diagenesis, Facies Examination, Paleocurrent Assessment.
10. **Geochemistry:** A branch of geology focused on the examination of the chemical composition of terrestrial materials and the chemical processes occurring inside Earth's systems. This discipline aids in comprehending elemental distribution and geochemical cycles. Principal Subjects: Elemental distribution, isotopic analysis, and geochemical cycles.
11. **Geophysics:** A branch of geology that applies physical concepts to investigate the Earth's morphology, gravitational and magnetic fields, interior composition, and processes. Methods include seismic waves, magnetic measures, and gravitational assessments. Principal Subjects: Seismology, geomagnetism, and gravitational research.
12. **Engineering Geology:** A branch of geology focused on the use of geological principles to address engineering challenges associated with building and land

utilization. It evaluates site conditions, stability, and hazards related to geological characteristics. Principal Subjects: Lithology and Mineralogy, Geological Formations, Soil Mechanics, and Dam Engineering.

13. **Environmental Geology:** A branch of geology that examines the connections between people and the geological environment, focusing on matters such as natural hazards, resource management, and the environmental consequences of geological processes. Principal Subjects: Environmental impact assessments, pollution mitigation, natural hazard management, slope stability, and foundation engineering.
14. **Remote Sensing and Geographic Information Systems (GIS):** A branch of geology that focuses on the analysis and interpretation of geological data and trends using satellite images and GIS technologies. Principal Subjects: Image analysis, geographic data management, and geospatial analysis.

Scopes of Geology

1. **Resource Exploration and Management:** Geology is essential for the identification and management of natural resources, including minerals, oil, natural gas, coal, and groundwater. It encompasses exploration methodologies, resource assessment, and sustainable stewardship.
2. **Environmental Protection:** Geologists evaluate environmental effects, oversee waste management, and formulate plans for pollution mitigation. They investigate soil and groundwater pollution and engage in remedial initiatives.
3. **Natural Hazard Assessment and Mitigation:** Geologists analyze and forecast natural hazards like earthquakes, volcanic eruptions, landslides, and floods. They focus on constructing structures and infrastructure capable of withstanding these threats and creating early warning systems.
4. **Engineering and Construction:** In civil engineering, geology is essential for site evaluation, foundation design, and structural stability. It aids in comprehending soil characteristics, rock mechanics, and groundwater conditions.
5. **Comprehending Earth's History and Evolution:** Geology offers insights into the planet's history via the examination of rock formations, fossils, and geological processes. This aids in the reconstruction of ancient ecosystems, climatic variations, and biological evolution.
6. **Planetary Geology:** Applies geological ideas to extraterrestrial bodies, facilitating the comprehension of their composition, surface processes, and historical development. This perspective is crucial for space research and comprehending the geology of celestial entities.
7. **Education and Research:** Geologists advance scientific inquiry and pedagogy, augmenting our understanding of terrestrial processes and systems. They are

employed in academic institutions, research organizations, and public scientific education.

Applications of Geology

1. **Exploration and Extraction of Natural Resources**

A. **Geological Prospecting**

a. Objective: Identify and assess reserves of precious minerals, including gold, silver, copper, and iron.

b. Methods: Geophysical surveys, geochemical analysis, drilling, and sampling.

c. Significance: Crucial for the mining sector, supplying raw materials for production and technology.

B. **Oil and Gas Exploration**

a. Objective: Identify prospective oil and gas fields and evaluate their economic feasibility.

b. Methods: Seismic surveys, well logging, and geological mapping.

c. Significance: Essential for the energy sector, guaranteeing a reliable supply of fossil fuels.

2. **Environmental Management**

A. **Contamination of Soil and Groundwater**

a. Objective: Evaluate and rectify polluted locations to safeguard human health and the environment.

b. Techniques: Soil sampling, groundwater monitoring, and risk evaluation.

c. Significance: Guarantees potable water, maintains healthy ecosystems, and adheres to environmental standards.

B. **Waste Management**

a. Objective: Design and administer waste disposal facilities to minimize environmental effects.

b. Techniques: Geotechnical investigations, site assessment, and surveillance.

c. Significance: Mitigates soil and water pollution, reduces health hazards, and fosters sustainable waste management methods.

3. **Evaluation and Mitigation of Natural Hazards**

A. **Seismic Prediction and Mitigation**

a. Objective: Comprehend and anticipate seismic phenomena to mitigate earthquake consequences.

b. Techniques: Seismology, fault mapping, and hazard modeling.

c. Significance: Improves public safety, directs building regulations, and informs emergency readiness.

B. **Evaluation of volcanic hazards**

a. Objective: Observe and forecast volcanic activity to reduce hazards.

b. Techniques: Volcanology, remote sensing, and geochemical analysis.
c. Significance: Safeguards communities, infrastructure, and air traffic against volcanic threats.

4. Engineering Geology

A. Construction Site Investigation

a. Objective: Evaluate geological conditions to ensure safe and stable building projects.
b. Techniques: Geotechnical drilling, soil analysis, and slope stability assessment.
c. Significance: Guarantees the safety and longevity of constructions like buildings, bridges, and dams.

B. Tunnel and Excavation Initiatives

a. Objective: Design and build tunnels and other subterranean constructions.
b. Techniques: Rock mechanics, hydrogeology, and geotechnical engineering.
c. Significance: Promotes infrastructure advancement, encompassing transportation and utilities.

5. Management of Water Resources

A. Subsurface Water Investigation and Administration

a. Objective: Identify and administer groundwater resources for sustainable use.
b. Techniques: Hydro geological studies, well drilling, and aquifer testing.
c. Significance: Guarantees a dependable water supply for consumption, agriculture, and industrial use.

B. Management of Surface Water

a. Objective: Administer Rivers, lakes, and reservoirs to fulfill human and ecological requirements.
b. Techniques: Hydrological modeling, watershed management, and flood risk evaluation.
c. Significance: Facilitates water provision, flood management, and ecological well-being.

6. Research on Climate Change

A. Paleoclimatology

a. Objective: Reconstruct historical climatic conditions with geological evidence.
b. Techniques: Analysis of ice cores, examinations of sediment cores, and examination of fossils.
c. Significance: Offers insights into natural climatic variability and enhances projections of future climate change.

B. Carbon Sequestration

a. Objective: Formulate techniques for the sequestration and storage of carbon dioxide to alleviate climate change.
b. Techniques: Geological sequestration, surveillance, and simulation.

c. Significance: Mitigates greenhouse gas emissions and facilitates the attainment of climate objectives.

7. Scholarly inquiry and instruction

A. Geological Investigation

a. Objective: Enhance scientific understanding of terrestrial dynamics and history.

b. Techniques: Field investigations, laboratory experiments, and theoretical modeling.

c. Significance: Facilitates scientific discovery, shapes public policy, and instructs future geologists.

B. Education and Public Engagement

a. Objective: Inform students and the public on geology and its significance.

b. Techniques: Instruction, public presentations, and instructional resources.

c. Significance: Enhances awareness of geological concerns, fosters scientific literacy, and motivates future generations.

8. Urban Planning and Development

A. Land Utilization Planning

a. Objective: Direct land development to guarantee sustainable and secure use.

b. Techniques: Geological mapping, risk assessment, and resource appraisal.

c. Significance: Facilitates sustainable development, mitigates natural disaster risks, and conserves resources.

B. Infrastructure Advancement

a. Objective: Develop and construct infrastructure that considers geological conditions.

b. Techniques: Geotechnical analysis, material evaluation, and environmental impact assessment.

c. Significance: Safeguards the integrity and durability of infrastructure, decreases maintenance expenditures, and preserves the environment.

The applications of geology include several domains, considerably aiding in natural resource management, environmental conservation, hazard mitigation, engineering, water resource management, climate change research, academic inquiry, education, urban planning, and infrastructure construction. The amalgamation of geological expertise and methodologies is crucial for tackling the most urgent concerns confronting humanity today.

Chapter 2
Introduction to Planetary Sciences

Planetary science, referred to as planetary geology or planetary astronomy, encompasses the study of planets, moons, and planetary systems, along with the processes that contribute to their formation and morphology. This multidisciplinary discipline integrates aspects of astronomy, geology, atmospheric science, chemistry, and physics to comprehend the nature and development of celestial bodies.

Key Areas of Study:

1. **Planetary Formation**: Planetary Formation examines the processes by which planets, moons, and smaller celestial bodies originate from gaseous and dusty accretion disks around nascent stars. It entails comprehending protoplanetary disks, planetesimals, and the accretion process that results in the creation of terrestrial planets, such as Earth, and gas giants, such as Jupiter.
2. **Planetary Interiors:** Planetary Interiors examines the composition, structure, and dynamics of planetary cores, mantles, and crusts. Methods include seismic investigations, gravity field assessments, and magnetic field evaluations to deduce internal characteristics.
3. **Planetary Surfaces:** This section examines surface features such as mountains, craters, volcanoes, and valleys on planets and moons. The processes that sculpt planetary surfaces include impact cratering, volcanism, erosion, and tectonics.

4. **Planetary Atmospheres:** This section examines the composition, structure, meteorological phenomena, and climatic systems of planetary atmospheres. The progression of planetary atmospheres aids scientists in comprehending phenomena such as greenhouse effects, atmospheric depletion, and prospective habitability.
5. **Moons and Rings:** This section concentrates on the natural satellites of planets, their creation, orbital dynamics, and surface geology. We examine planetary rings, such as Saturn's, to understand their origins and development.
6. **Asteroids, Comets, and Small Bodies:** This section examines small celestial entities within the solar system, such as asteroids in the asteroid belt and comets that originate from the Kuiper Belt and Oort cloud. These artifacts provide understanding of the primordial solar system and planetary development.
7. **Exoplanets:** Scientists investigate planets outside our solar system (exoplanets) using diverse detection techniques such as the transit method and radial velocity. The examination of exoplanets aids scientists in comprehending the variety of planetary systems and the possibility for extraterrestrial life in the cosmos.

Techniques and instruments:

1. **Telescopes:** Both ground-based and space-based telescopes provide observational data on planetary systems and exoplanets.
2. **Space Missions:** Robotic spacecraft, like NASA's Voyager, Cassini, Curiosity, and ESA's Rosetta, investigate planets and moons, delivering detailed observations.
3. **Remote Sensing and Spectroscopy:** Methodologies used to investigate planetary atmospheres and surface compositions via the analysis of light emitted or reflected by celestial bodies.
4. **Laboratory Simulations:** Researchers replicate planetary circumstances in laboratories to comprehend the mechanisms of surface erosion and chemical reactions on various celestial bodies.
5. **Computer Modeling**: Computational simulations replicate the genesis of planetary systems, climate dynamics, and surface interactions.

What is the significance of studying planetary science?

1. **Comprehending Earth:** Through the examination of other planets, researchers get knowledge about Earth's processes, facilitating an understanding of its climate, geology, and prospective hazards such as asteroid collisions.
2. **Search for Life:** Examining conditions on extraterrestrial planets and moons aids in pinpointing locations where life may now present or may have previously lived.
3. **Exploration and Discovery:** Planetary science propels exploration efforts, enhancing human understanding of our solar system and beyond.

Planetary Science inside the Solar System:

1. **Terrestrial Planets:** Mercury, Venus, Earth, and Mars are solid, rocky entities with definitive surfaces.
2. **Gas Giants:** Jupiter and Saturn are colossal planets characterized by dense atmospheres mostly composed of hydrogen and helium.
3. **Ice Giants:** Uranus and Neptune largely consist of ice, including water, ammonia, and methane.
4. **Dwarf Planets:** Pluto, Eris, and further remote entities inside the Kuiper Belt.

This domain persists in expanding due to novel findings from missions, telescopic observations, and technological breakthroughs, all contributing to the elucidation of the enigmas surrounding the planets inside our solar system and beyond.

Chapter 3
Introduction to Fundamental Principles of the Universe's Origin

For centuries, humans have been fascinated by the genesis of the universe, which combines aspects of science, philosophy, and religion. Over time, research in science has developed a structure for understanding the origins and evolution of the universe. An overview of the fundamental ideas behind the universe's genesis is provided here:

1. **The Big Bang Theory:** The most commonly recognized scientific notion about the universe's beginning is the Big Bang Theory. This theory suggests that the cosmos has been expanding ever since it formed around 13.8 billion years ago as an extremely hot and dense spot. The first evidence supporting the Big Bang Theory is Nucleosynthesis suggests that hydrogen, helium, and other light elements are expected to be abundant. In the 1920s, the second evidence supporting the Big Bang Theory is Edwin Hubble recorded a finding that galaxies were moving apart, which supports the Big Bang theory by suggesting that the universe is expanding. The third Evidence Supporting the Big Bang Theory is the Cosmic Microwave Background Radiation (CMB).The remnants of the Big Bang explosion are known as the Cosmic Microwave Background Radiation (CMB), which has been discovered in 1965 and provides a picture of the early cosmos.

Figure 3.1: The figure representing the Big Bang, capturing the explosion of light and energy that marked the beginning of the universe.

2. **Cosmic Inflation:**According to the hypothesis of cosmic inflation, the cosmos expanded very quickly in the first microscopic fraction of a second shortly after the Big Bang. The flatness and horizon concerns, among other issues with the Big Bang Theory, are resolved, and it explains why the cosmos appears uniform in all directions. On a wide scale, the cosmos seems uniform and isotropic, which may be explained by inflation. The genesis of the very initial density fluctuations that resulted in the universe's large-scale structure may be explained by cosmic inflation. Inflationary model predictions are consistent with the distribution of galaxies and cosmic structures. Inflation is supported by the Cosmic Microwave Background Radiation's (CMB) uniform temperature across the sky.

Figure 3.2: The image representing Nucleosynthesis in the first few minutes after the Big Bang, illustrating the creation of hydrogen and helium through intense nuclear reactions.

3. **General Relativity and the Role of Gravity:** Our knowledge of gravity was completely transformed by Albert Einstein's general relativity theory, which was

developed in 1915 and defined gravity as the curvature of spacetime created by mass and energy. General relativity demonstrates that enormous objects like stars and planets create a curvature in spacetime that affects the movement of other objects, in contrast to Newton's theory of gravity as a force functioning at a distance. The mechanics of the universe's expansion and the creation of massive cosmic structures like galaxies and galaxy clusters can no longer be explained without this revolutionary idea. In addition, general relativity predicts the existence of gravitational waves, which are ripples in spacetime caused by immense cosmic events like merging black holes, and black holes, which are areas of spacetime with gravity so strong that not even light can escape. The crucial role that general relativity plays in modern cosmology has been further established by the observation of both of these predictions: black holes have been observed and gravitational waves have been spotted up by observatories such as Virgo and LIGO. This theory's applications have given us a deep framework that influences how we see the universe and encourages further research into the nature of gravity, time, and space.

4. **Formation of Matter and Structure:** After the cosmos first expanded, temperatures decreased, which made it possible for particles to mix and create atoms. At this point, gravity was essential because it slowly brought these atoms together to create stars, galaxies, and ultimately gigantic structures like clusters of galaxies. Around 380,000 years after the Big Bang, recombination was a crucial process throughout this time. Protons and electrons were able to mix to create neutral hydrogen atoms as the cosmos cooled further. Due to the dissociation of matter and radiation caused by this event, the universe became transparent, and the cosmic microwave background radiation was emitted, a faint light that still exists today and provides an overview of the early universe. Nucleosynthesis, which occurred in the first minutes after the Big Bang, was another crucial activity. In this short time, the universe's initial light elements mostly hydrogen and helium as well as trace quantities of lithium were created by nuclear process. With the passage of time, gravitational collapse where areas with marginally greater densities attracted more matter was created by tiny density changes in the cosmos. As a result of gravity gradually pulling gas and dust together, the first stars and galaxies were formed. These processes drove the development of more complex structures over millions of years, resulting in the enormous and complex cosmic web of clusters and galaxies that we see in the universe today.
5. **Dark Matter and Dark Energy:** Despite the fact that they are two of the most enigmatic components of our universe, dark matter and dark energy play absolutely essential functions in the structure and development of the universe. Due to the fact that it does not emit, absorb, or reflect light, dark matter, which accounts for about

27% of the total mass-energy capacity of the universe, is not visible to the naked eye. On the other hand, the gravitational impact of this object is absolutely necessary for the development of galaxy clusters and galaxies.

Dark matter functions as a cosmic scaffold, bringing about the formation of the large-scale structure of the universe. It does this by providing the gravitational attraction that is necessary to keep galaxies together. Even though the precise form of dark matter is still unresolved there is evidence from observations that supports its existence. These observations include the rotation rates of galaxies and the bending of light that occurs as a result of gravitational lensing. Dark energy, on the other hand, is a force that is even more mysterious than light energy and is responsible for about 68% of the total energy in the universe. It would seem that dark energy, as opposed to dark matter, is the driving force behind the rapid expansion of the universe. This means that galaxies are moving apart from each other at a rapid rate. This rapid expansion was first discovered by supernova observations in the latter part of the 20th century. As a result, scientists have hypothesised the existence of dark energy, despite the fact that the nature of dark energy's underlying components continues to be one of the most perplexing puzzles in the field of cosmology. Despite the fact that dark matter and dark energy together constitute around 95% of the entire substance of the universe, the qualities of these two components continue to be a source of consternation for scientists and motivate fresh investigations into the underlying processes that govern the universe.

6. **Philosophical and Theological Perspectives:** When it comes to the question of how the cosmos came into being, philosophical and theological viewpoints provide a variety of various interpretations that are based on philosophical and religious traditions. Among these viewpoints is the creationist viewpoint, which asserts that a divine person or a higher force was responsible for the creation of the world and life. This perspective is prevalent in a number of different theological traditions, each of which has its own unique narratives about the nature of creation and the purpose of a creator. Several monotheistic faiths, for instance, have the belief that God is the one who purposely created the cosmos, therefore establishing a spiritual basis for existence and the existence of man. On the other hand, ideologies such as deism and theism provide more complex perspectives on the connection that exists between a god and the cosmos. The theory of deism believes that a god or gods created the cosmos, but that they do not interfere with its functioning. This enables the universe to operate independently in accordance with the rules of nature. There were certain Enlightenment intellectuals who held the belief that the nature of God's creation might be uncovered through the use of reason and the observation of the natural world. This viewpoint was widespread among these individuals. Theism, on the other

hand, is characterized by the belief in a personal God who not only created the world but may also continue to engage with it, although there are a variety of interpretations of this concept. When taken as a whole, these viewpoints provide a variety of insights into the beginnings and operations of the cosmos. They do so by situating it within a framework that places an emphasis on either the existence or absence of divine activity in the natural order.

7. **Multiverse Hypothesis**:According to the multiverse theory, our world might be only one of many parallel universes, each of which has its own set of physical rules and constants. Despite the fact that it is a theoretical theory, the multiverse hypothesis originates from certain interpretations of quantum physics and cosmological models.

Figure 3.4: The image representing the multiverse hypothesis, with multiple bubble-like universes floating in a vast cosmic void, each with its own distinct look.

8. **Current Research and Future Directions:** Research that is now being conducted on the origin and nature of the universe is making progress via cutting-edge areas such as observational astronomy and particle physics. This is opening up new opportunities for comprehending the mysteries of the universe. Scientific tools such as the James Webb Space Telescope (JWST) have brought about a revolution in the field of observational astronomy. These devices have enabled researchers to see

farther into space and time than they ever have before. Through the collection of light from the earliest galaxies, the James Webb Space Telescope (JWST) and other future observatories intend to offer information that has never been seen before on the genesis and development of cosmic structures a short time after the Big Bang. The nature of dark matter, the creation of galaxies, and the early stages of star growth are all important problems that may be answered with the assistance of scientific evidence. At the same time, research in particle physics is being conducted at facilities such as the Large Hadron Collider (LHC), which is investigating basic particles and forces in order to investigate the fundamental building blocks of the cosmos. It is possible for scientists to mimic circumstances that were comparable to those that existed only a few minutes after the Big Bang by accelerating particles to speeds close to the speed of light and monitoring the collisions that arise from this acceleration. In order to get a better knowledge of how particles acquired mass, the nature of the forces that produced the universe, and the possibility of new particles that might disclose previously unknown features of the universe, this study is very important. The future of cosmology is being shaped by the combined efforts of advancements in observational astronomy and particle physics. These approaches are providing discoveries that have the potential to reframe our knowledge of the basic principles that govern the universe and its beginnings.

In order to get a comprehensive and continuous understanding of the genesis of the universe, a scientific determination that integrates observational astronomy, theoretical physics, and philosophical inquiry is being undertaken. However, research continues to enhance our knowledge and address the numerous problems that remain unsolved concerning the origin of the universe with regard to the Big Bang Theory, which continues to be the most commonly accepted explanation. The cornerstone of our knowledge of the genesis and development of the universe is comprised of these ideas, despite the fact that the discipline is always developing with the introduction of new findings and hypotheses.

Chapter 4
The Origin of the Solar System

Introduction: Over 4.6 billion years ago, our solar system was formed, and its history is a great story that dates back to the beginning of time.Astronomical data, geological evidence, and theoretical models have been used to put together a story that has been attached together. Not only is it a trip into the processes that create planetary systems across the cosmos, but it is also a search to understand where we originate from, which is why it is important to comprehend the beginning of the Solar System. The most prominent hypotheses and pieces of evidence about the origin and early development of our Solar System are discussed in depth in this chapter.

Some of the most prominent concepts and hypotheses that have been put forth by a variety of scientists in an effort to explain the formation of our solar system are as follows:

Early Theories

- **Descartes' Vortex Theory (1644):**It was proposed by René Descartes in 1644 that the solar system was generated from a rotating vortex of material. This hypothesis was known as the Descartes' Vortex hypothesis.
- **Buffon's Comet Collision Theory (1745):**Comte de Buffon, Georges-Louis Leclerc, put forward the hypothesis that the planets were formed when a comet collapsed with the Sun, causing material to be ejected and then condense into the planets.

Nebular Hypothesis

- **Kant's Nebular Hypothesis (1755):**Immanuel Kant put forward the hypothesis that the solar system originated from a cloud of gas and dust that rotated around itself.
- **Laplace's Nebular Hypothesis (1796):**Kant's hypothesis was further expanded upon by Pierre-Simon Laplace, who proposed that the planets were formed as a result of the nebula progressively contracting and cooling.

Encounter Hypotheses

- **Chamberlin-Moulton Planetesimal Hypothesis (1905):** Thomas Chamberlin, an American geologist, and Forest Moulton, an astronomer, came up with the Chamberlin-Moulton Planetesimals Hypothesis in 1905. This hypothesis proposes that the solar system was created from planetesimals, which are tiny solid particles, rather than gas. This idea proposes that a passing star got near to the Sun, which resulted in the separation of a stream of material owing to the gravitational forces that were there. During the cooling process, this material broke up into planetesimals, which are little solid bodies. The planets and other things in the solar system were progressively formed as a result of the collisions and coalescence of planetesimals over the course of time.
- **Gaseous Tidal Hypothesis Jeans & Jeffreys (1925):** The Gaseous Tidal Hypothesis, which was first out by James Jeans and Harold Jeffreys in the early 20th century, proposes that the solar system was created as a consequence of a near-collision between the Sun and a massive star that was passing by at the time. According to this idea, the powerful tidal forces that were exerted by the passing star were responsible for the removal of gaseous material from the surface of the Sun. The planets were formed as a result of the eliminated gas eventually cooling and condensing. The denser areas of the material that was ejected were responsible for the formation of the bigger gas giants, while the less dense regions were responsible for the formation of the smaller, rocky planets.
- **Lyttleton binary star Hypothesis (1938):**An astronomer from the United Kingdom named Raymond Lyttleton put up the Lyttleton double Star Hypothesis in the middle of the 20th century. This hypothesis proposes that the Sun was once a component of a double star system. A companion star to the Sun is said to have ultimately come extremely close to a huge third star, as suggested by this hypothesis. The companion star became cracked and broken apart as a result of the gravitational interaction that occurred between these two stars. Following the formation of a disk of material around the Sun, the debris that resulted from this disruption gradually merged together to create the planets and other entities that make up the solar system.

Other Theories

- **Accretion Theory of Carl von Weizsäcker (1944 to 1948):** Carl von Weizsäcker has been recognized as being the person who most significantly created and enlarged upon the Accretion Theory in the middle of the 20th century. This theory suggests that planets evolved gradually via the accretion of smaller particles. According to this idea, the process of planetary formation starts with a disk of gas and dust encircling a young star. This disk has the potential to become a planet. The smaller particles that are included inside this disk come into contact with one another and attach to one another, resulting in the formation of bigger and larger entities throughout the course of time. Eventually, these entities, which are known as planetesimals, will develop into full-sized planets because they continue to expand by attracting more material via the gravitational forces. The Accretion Theory provides an explanation for the creation of planetary systems that is generally accepted. It is a vital component of the Nebular Hypothesis, which is a more comprehensive theory.
- **Captures hypothesis of Michael Mark Woolfson (1964):** In 1964, Michael Mark Woolfson put up the Capture Hypothesis, which hypothesizes that the planets in the solar system did not originate from the Sun but rather originated from another location in space. In accordance with this idea, a protostar or planetary system that was traveling near to the Sun was able to catch some of the planets into its orbit using the gravitational attraction of the Sun.

Modern Variations

- **Protoplanetary Disk Model:**Alastair G. W. Cameron and William W. Fowler made key contributions to the development of the Protoplanetary Disk Model in the 1970s. This model is a contemporary approach to the Nebular Hypothesis. Using a disk of gas and dust, also known as the protoplanetary disk, which surrounds a newborn star, this model improves upon the Nebular Hypothesis by concentrating on the creation of planets from the protoplanetary disk. This concept proposes that the material of the disk eventually collects into planetesimals and subsequently into protoplanets via processes of accretion and gravitational interactions, which ultimately results in the development of planets and other bodies in the solar system.
- **Core Accretion Model:**Geoffrey W. Marcy and William K. Hartmann were the primary contributors to the development of the Core Accretion Model in the 1980s. This model proposes that rocky or ice cores formed first, and subsequently they attracted gas to become gas giant planets. This concept provides an explanation for the development of gas giants like as Jupiter and Saturn by describing the initial accumulation of solid cores together with the subsequent accumulation of gas from the protoplanetary disk in the surrounding region.

- **Gravitational Instability Model:**One of the most notable contributions that A. P. Boss made in the late 1990s was the development of the Gravitational Instability Model. This model suggests that gas giant planets developed directly from the protoplanetary disk via gravitational instability. This hypothesis proposes that some portions of the protoplanetary disk were gravitationally unstable and collapsed straight to produce gas giant planets. This model eliminates the need that solid cores initially develop. A different approach to the Core Accretion concept is offered by this concept, which places a higher priority on the part that disk instability plays in the development of gas giants.

Note: It is important to note that while the nebular hypothesis is the explanation that is most largely accepted, there continues to be study and discussion over the particular aspects of the birth of the solar system.

The Nebular Hypothesis

A. **Origin and Proponents**: The Nebular Hypothesis is the concept that has received the greatest support from the scientific community about the creation of the Solar System. In the year 1755, the German philosopher Immanuel Kant was the first to suggest the idea, and in the year 1796, the French mathematician Pierre-Simon Laplace was the one who advanced it further.

B. **Basic Premise**: The Solar System originated from a massive cloud of interstellar gas and dust that rotated around itself and was referred to as a solar nebula. There is a possibility that the collapse of the nebula was caused by external factors, such as a nearby supernova explosion, which caused it to collapse under its own gravity.

C. **Formation of the Protosun**: The material that would eventually form the protosun, the earliest forerunner to our Sun, collected near the center of this rotating disk, which was the source of the material. The core temperatures and pressures of the protosun rose as it expanded in mass, finally reaching the point where nuclear fusion could spark, which resulted in the formation of the Sun when it had reached the limit of fusion.

D. **Formation of the Protoplanetary Disk**: It is believed that the planets, moons, asteroids, and comets all originated from the protoplanetary disk, which is a disk of gas and dust that surrounds the planet. These heavier components had been created in previous generations of stars and spread into space by supernovae. This disk was mostly made of hydrogen and helium, with trace quantities of heavier elements.

E. **Temperature Gradient**: There was a large difference in temperature between the inner and outer portions of the disk, with the inner areas being much hotter than the outside regions. When it came to figuring out the make-up of the planets, this temperature differential was an extremely important component. The only substances that could condense into solid particles in the inner, hotter areas were those that had high melting points, such as metals and silicates. There is a possibility that volatile molecules like as water, ammonia, and methane might freeze at the temperatures that are lower in the outer regions.

F. Accretion and Formation of Planetary Bodies

Accretion Process: A process known as accretion is the cause for the formation of planets from the protoplanetary disk. In the beginning, the disk included minute particles of dust and ice that hit with one another and fused to one another, resulting in the formation of bigger and larger aggregates. Planetesimals, which were bodies that were a few kilometers wide and had sufficient gravitational pull to attract additional material, were formed throughout the course of time by these aggregates.

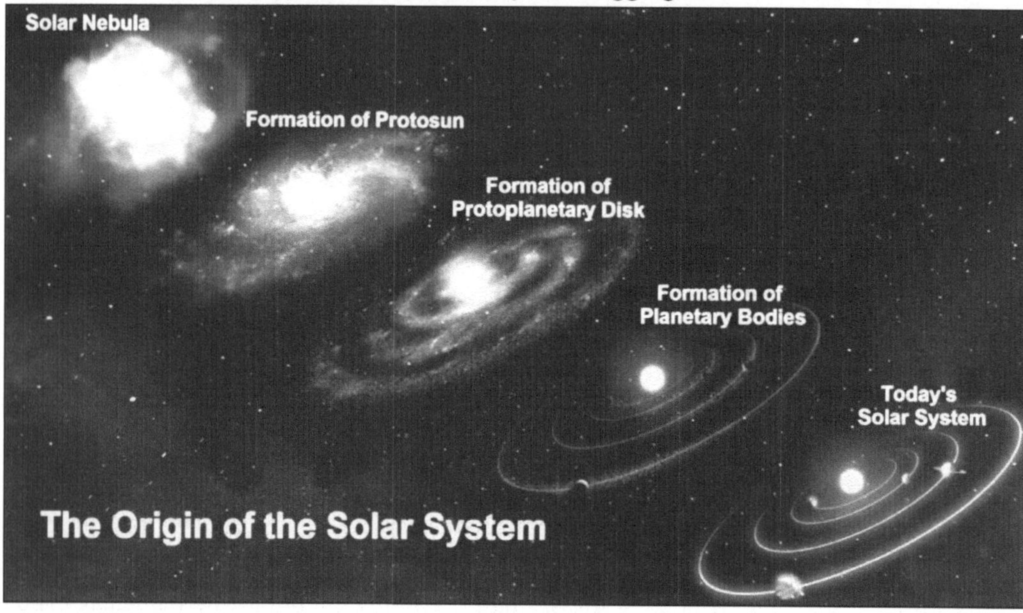

Figure 4.1:Diagram illustrating the Origin of the Solar System (Reference - Lincoln Learning Solutions, YouTube)

Terrestrial Planets: Metal and rock were the primary components of planetesimals that were found in the inner region of the Solar System, which was characterized by high temperatures. Eventually, these planetesimals merged together to form the terrestrial planets, which are Mercury, Venus, Earth, and Mars by their names. The fact that these planets are considerably smaller in size in comparison to the gas giants is a reflection of the limited availability of solid material in the portions of the protoplanetary disk that are significantly hotter.

Gas Giants and Ice Giants: After cross the "frost line," planetesimals have the potential to accumulate more material, which may include ices. Because of their greater masses, these planetesimals were able to extract hydrogen and helium from the disk that around them, which resulted in the formation of gas giants like as Jupiter and Saturn. The formation of ice giants, like as Uranus and Neptune, occurred farther out. These ice giants were predominantly made up of water, ammonia, and methane ices.

G. The Role of Planetary Migration and Gravitational Interactions

Planetary Migration: The significance of gravitational interactions and planetary migration in the formation of the early Solar System has been brought to light by recent study. While the big planets were forming, they did not stay in the orbits that they had first formed. On the contrary, they migrated through the disk, influenced by gravitational interactions with the surrounding gas and other planetesimals.

Nice Model: One notable event that most likely took place during this time period is referred to as the Nice Model, which was called after the city in France where it was created. According to this scenario, the large planets first formed in a configuration that was more compact, and then they migrated to their present locations throughout time. The Late Heavy Bombardment was a period in which the inner planets experienced an increase in the number of collisions by asteroids and comets. This migration might have had potentially catastrophic implications on the remainder of the Solar System, including the possibility of dispersing smaller bodies and contributing to the Late Heavy Bombardment.

Gravitational Interactions: The ultimate creation of the Solar System was influenced by the interactions between planets and Planetesimals, which led to the dispersion and redistribution of smaller objects.

H. Formation of Moons, Asteroids, and Comets: At the same time that the planets were coming into being, smaller bodies were simultaneously coming together all around the protoplanetary disk. In the far reaches of space, beyond the orbit of Neptune, a vast number of frozen bodies came into being, which eventually led to the formation of the Kuiper Belt. While some of these objects, such as Pluto and Eris, are of a size that allows them to be categorized as dwarf planets, others continue to possess the characteristics of comets. Only few of the planetesimals that existed in the inner Solar System were able to be integrated into planets. The majority of them formed into the asteroids that we see today, which are largely situated in the asteroid belt that lies between Mars and Jupiter. Many of them remained as smaller objects. It is thought that the gravitational force of Jupiter had a role in preventing these planetesimals from becoming a planet of their own. The development of moons is a complicated process that most likely included a number of distinct factors. Some moons, such as the moon that orbits the Earth, are theorized to have originated from massive collisions. The most prominent explanation about the creation of the Moon proposes that a planet of the size of Mars collided with the early Earth, resulting in the ejection of material that eventually merged into the Moon. There is a possibility that other moons, notably those of the gas giants, come into being in situ, inside little disks of material that encircle the planets that are expanding. In addition, there is the possibility that some of them are captured objects, such as the moon Triton of Neptune, which is believed to be a Kuiper Belt object that has been caught.

I. **Clearing of the Nebula and the Final Stages:** As soon as the Sun achieved its maximum mass and started the process of stable hydrogen fusion, the solar wind, which is a stream of charged particles that the Sun emits, became powerful enough to sweep away the gas that remained in the protoplanetary disk. The primary phase of planet formation came to an end at this point, since the leftover material was either expelled from the Solar System or absorbed into objects that were already in existence. The planets began to settle into the orbits that we see today during this time period, which also indicated the beginning of a more stable age for the Solar System. In spite of this, the early Solar System was nevertheless a dynamic environment, with regular collisions between planetesimals. These impacts resulted in the cratering of planetary surfaces and the transport of water and organic compounds to the inner planets, including Earth. The dynamic interaction of gravity, chemistry, and the dynamics of a protoplanetary disk was the driving force behind the creation of the Solar System, which was a complicated and chaotic process. It is a monument to the remarkable processes that rule the universe that the Solar System's creation narrative is a story that begins with the collapse of a molecular cloud and ends with the birth of a vast variety of planets, moons, and smaller objects.

Having an understanding of this process not only provides us with a better understanding of our own beginnings, but it also assists us in our quest for other planetary systems and planets out in the galaxy that may or may not be habitable. The story of how the Solar System came into existence continues to be an essential component of our comprehension of the processes that lead to the development of planets and the factors that are necessary for the existence of life.

Chapter 5
Meteorites and Impact Craters

Introduction

A significant amount of information on the history of the solar system, the surfaces of the planets, and the processes that occur in the cosmos may be obtained from meteorites and impact craters. Meteorites are pieces of asteroids, comets, or even other planets that survive their voyage through the atmosphere of Earth and reach the surface of the earth. Impact craters, which are generated when such things collide with the surfaces of planets, provide a record of the dynamic history of the solar system as well as the role that collisions have played in sculpting the landscapes of the cosmos. The nature of meteorites, their classification, the development of impact craters, and the features of impact craters are all topics that are covered in this chapter.

Meteorites

What Are Meteorites?

A meteorite is a solid piece that originates from space passes it through the atmosphere of Earth and ultimately lands on the surface of the planet. Asteroids in the asteroid belt are the source of the vast majority of meteorites; nevertheless, some meteorites originate from comets, the Moon, and even Mars. When the object enters the atmosphere of Earth, the friction causes it to get heated, which causes it to light brilliantly like a meteor or fireball during its journey. It is considered a meteorite if it is able to survive its journey into the atmosphere and then arrive on the surface of the Earth. Scientists are able to investigate the earliest solar system, the creation of planets, and the material composition of asteroids, moons, and planets because of the direct samples of other solar system bodies that meteorites provide. This is one of the reasons why meteorites are very important.

Classifications of Meteorite

Meteorites are generally classified into three broad categories based on their composition: stony meteorites, iron meteorites, and stony-iron meteorites.

1. **Iron Meteorites**: Iron meteorites are primarily made up of iron-nickel alloys, and they originate from the cores of differentiated asteroids. Because of the passage of time, heavy metals like iron and nickel eventually made their way to the center of these bodies, resulting in the formation of a metallic core. Iron meteorites often display a characteristic pattern known as the Widmanstätten pattern, visible when they are cut and polished. This pattern is formed by the slow cooling of iron and nickel crystals over millions of years.

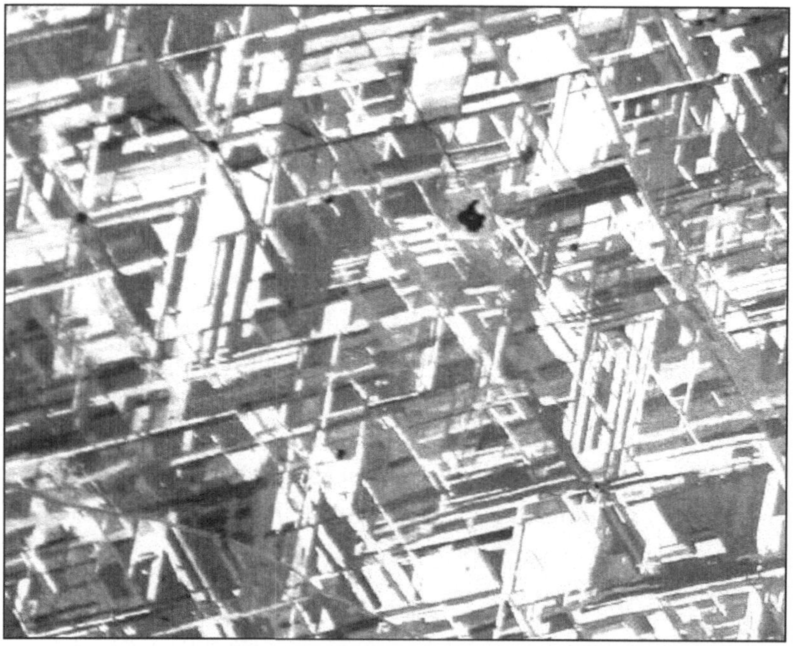

Figure 5.1: Widmanstätten pattern in Iron Meteorites

2. **Stony Meteorites (Chondrites and Achondrites)**: Chondritic inclusions, also known as chondrules, are comprised of tiny spherical inclusions and are the most common form. These silicate-rich, quickly cooled molten droplets, known as chondrules, provide a glimpse into the events that occurred in the early stages of the solar system. Since their genesis 4.6 billion years ago, chondrites have remained

virtually unchanged, which is one of the reasons why they are regarded to be among the most primordial minerals. In the absence of chondrules, stony meteorites are referred to as achondrites. They have experienced major transformations on their parent bodies, such as the Moon, Mars, or asteroids like Vesta, via the processes of melting and differentiation. There is a striking similarity in composition between these meteorites and terrestrial igneous rocks.

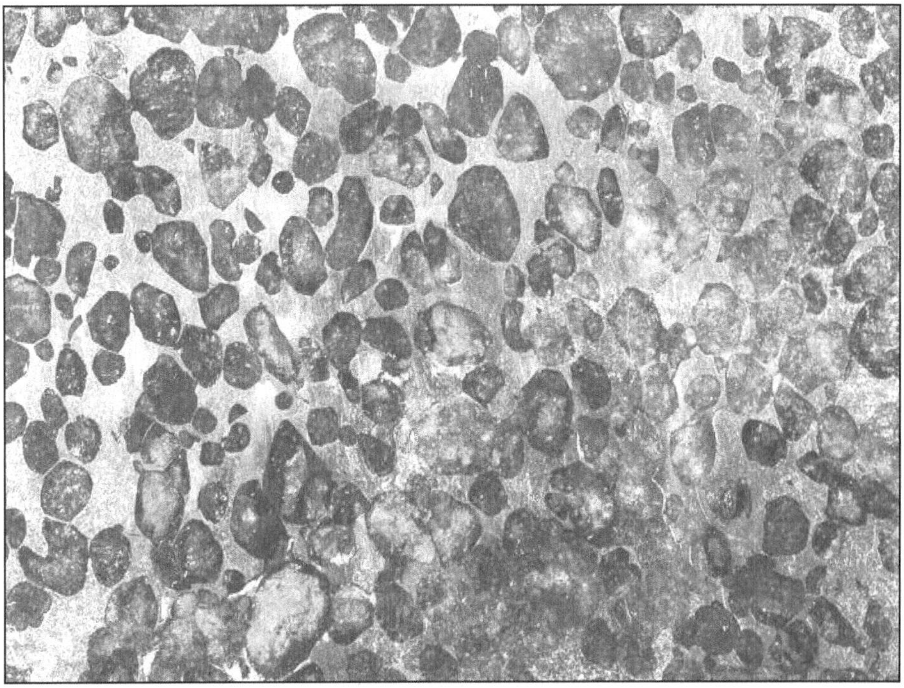

Figure 5.2:Pallasites Stony-Iron Meteorites contain olivine crystals and iron-nickel metal.

3. **Stony-Iron Meteorites**: An almost equal proportion of silicate minerals and iron-nickel metal may be found in these extremely rare meteorites." One of the most well-known types is called pallasites, and it is characterized by olivine crystals that are embedded in a metallic matrix. Materials from the border between the core and the mantle of differentiated asteroids are assumed to be represented by these objects.

Origins and Delivery to Earth

The asteroid belt that lies between Mars and Jupiter is the source of the majority of meteorites. It is possible for sources to be ejected into orbits that ultimately coincide with the orbit of the Earth either as a result of collisions with asteroids or gravitational disturbances from Jupiter. There are meteorites that come from the Moon and Mars. These meteorites were driven off the surface of the Moon and Mars by significant impacts, and they were finally trapped by the gravity of Earth. Comets are infrequent sources of meteorites. Because of the high concentration of volatile components in them, they are often more brittle and have a lower probability of surviving entrance through the atmosphere. In addition to assisting us in comprehending the makeup of foreign celestial bodies, the study of meteorites additionally allows us to get an understanding of the processes of planetary differentiation and the development of the solar system.

Impact Craters

Formation of Impact Craters

The collision of a meteoroid, asteroid, or comet with the surface of a planet at a high velocity results in the formation of impact craters. These impacts release large quantities of energy, which results in the production of shock waves causing the surface to be deformed and excavated.

The process of crater formation occurs in three stages:

1. **Contact and Compression**: The meteoroid's kinetic energy is passed to the surface when it hits, making a shock wave that moves outward and downward. At the impact site, the temperature and pressure can be several million times higher than Earth's air pressure. This can turn the meteoroid and some of the target material into a vapor.
2. **Excavation**: The shock wave causes material to be removed from the impact site, resulting in the formation of a temporary crater that is far bigger than the impactor itself. The material that is released from the crater has the potential to travel large distances, resulting in the formation of a debris blanket surrounding the impact site. Some of the debris that is ejected may completely escapes the gravitational pull of a planet, while other fragments can re-enter the atmosphere and fall back as secondary meteorites. This difference is determined by the magnitude of the impact.
3. **Modification**:Especially in the case of more significant impacts, the transitory crater may collapse under the force of gravity after the first shock period. The formation of complex structures, central peaks, and terraced rims may occur in craters that are greater than four kilometers in diameter. The ultimate shape of the crater is formed as a result of this process. Some craters may be altered or obscured over the course of time as a result of processes such as erosion, tectonic activity, or volcanic activity; nevertheless, there are craters that have been preserved for billions of years, especially on geologically inert worlds such as the Moon and Mars.

Types of Craters

Impact craters vary in size and complexity depending on the energy of the impact, the composition of the impactor and target material, and the gravity of the planet or moon.

1. **Simple Craters**: Small impacts produce bowl-shaped craters that are basic in appearance and have elevated rims and smooth interiors. In most cases, these craters are formed when the impactor has a diameter of less than four kilometers. For instance, the well-known Barringer Crater in Arizona is a good illustration.

Figure 5.3: Barringer Crater in Arizona.

2. **Complex Craters**: Those with a greater magnitude of impact produce craters that are more intricate, with center peaks or rings, terraced walls, and flat floors. The formation of these craters occurs when the force of the impact is sufficient to cause the sides of the crater to collapse inward, resulting in the formation of a peak or ring of material that has been raised in the center of the crater. In addition to having a central peak and many rings, the Chicxulub Crater is a complex crater that is over 180 kilometers in diameter. It is the crater that is responsible for the global extinction event that wiped out the dinosaurs.

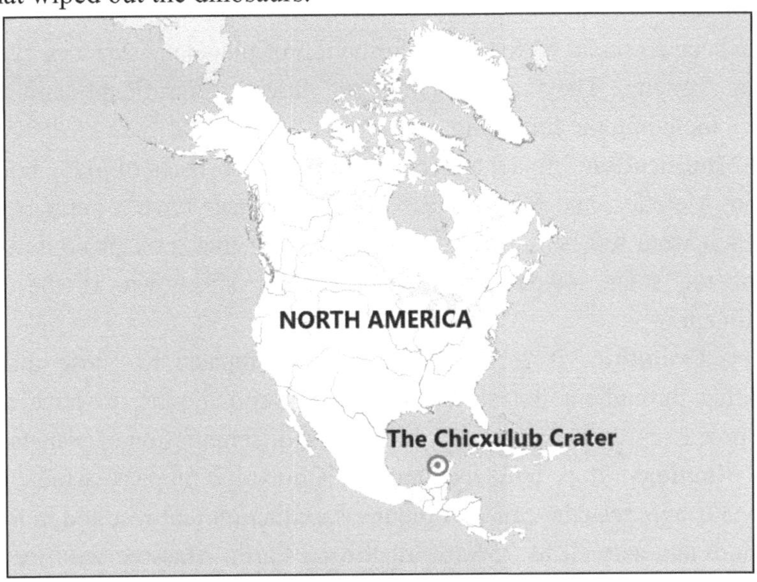

Figure 5.4: Present time location of The Chicxulub Crater, responsible for the mass extinction event that killed off the dinosaurs, is a complex crater over 180 km in diameter.

3. **Multi-Ring Basins**: Larger impacts result in the formation of multi-ring basins, which are characterized by the formation of concentric rings of mountains by the shock wave that reverberates across the surface of the planet. In most cases, the diameter of these constructions ranges from between hundreds and thousands of kilometers. One example of a multi-ring basin is the Orientale Basin on the Moon, which is considered to be one of the regions with the finest preservation.

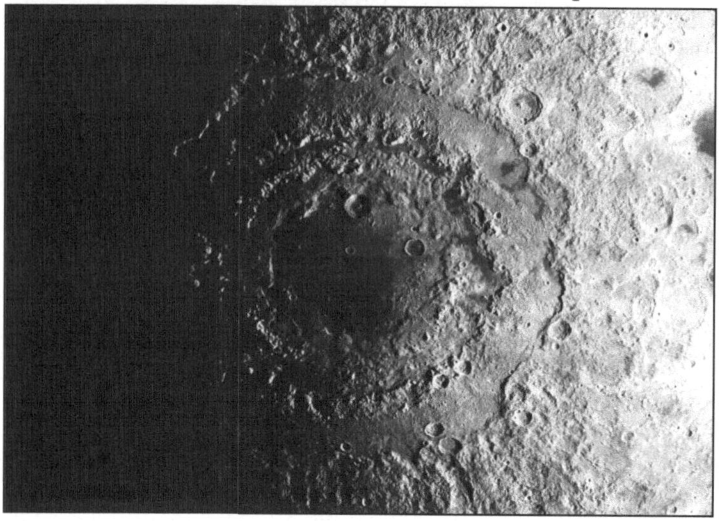

Figure 5.5: Basin of Mare Orientale on the Moon. Imaged by Lunar Orbiter 4 (1967) from altitude 2773 km. (Reference: Wikipedia, *the Free Encyclopedia.*)

Effects of Impacts

Impacts have been a crucial factor in the formation of planetary surfaces over the whole of the solar system. They are capable of having significant and long-lasting consequences, including the following:

1. **Surface Modification**: Impact craters dominate the surfaces of many bodies, such as the Moon, Mercury, and Mars. These ancient surfaces have a permanent record of impacts that were left behind by the bombardment that took place during the early stages of the solar system. This bombardment is known as the Late Heavy Bombardment.
2. **Planetary Evolution**: It is possible for large impacts to cause material to be redistributed throughout the surface of planets and moons, to have an effect on tectonic processes, and even to contribute to the differentiation of planets.
3. **Mass Extinctions**: It is believed that the Chicxulub impact, which occurred 66 million years ago, was the cause of the mass extinction that resulted in the extinction of the dinosaurs as well as 75% of all life on Earth. Massive wildfires, a "nuclear winter" effect caused by debris blocking sunlight, and acid rain were some of the worldwide repercussions that were caused by this catastrophe.

4. **Creation of Moons**: According to the huge impact theory, the Moon of Earth was created from the debris of a massive collision that occurred around the beginning of the history of the solar system between Earth and a planet that was about the size of Mars.

Famous Impact Craters

1. Barringer Crater, located in Arizona, is a small crater that measures 1.2 kilometers in width and 170 meters in depth. It was formed around 50,000 years ago by a tiny iron-nickel meteorite.
2. Chicxulub Crater, located on the Yucatán Peninsula, is an impact formation that spans 180 kilometers and was formed 66 million years ago and associated with the mass extinction of the dinosaurs.
3. Approximately two billion years ago, the Vredefort Crater in South Africa was formed and it is the biggest verified impact structure on Earth. Its original diameter is believed to have been over 300 kilometers.
4. Tycho Crater, located on the Moon, is a noticeable lunar crater that has brilliant beams of ejecta that reach over the surface of the Moon. This crater was generated by an impact that occurred relatively recently, around 108 million years ago.

A glimpse into the violent and dynamic processes that have built our solar system may be gained via the examination of meteorites and impact craters. Through the study of these occurrences, scientists are able to obtain a greater understanding of the creation of planets, the development of their surfaces, and even the possible dangers that may be presented by future impacts on Earth. Impact craters show the destructive force of collisions that have affected planetary landscapes for billions of years, while meteorites give samples from faraway worlds, which helps to untangle the composition and history of asteroids, moons, and planets. Meteorites also provide samples from solar systems that are very far away.

Chapter 6
The Solar System and their planets

The Solar System consists of the Sun and the celestial bodies that orbit it, including the eight major planets, dwarf planets, moons, asteroids, comets, and other objects. Here's an overview of the Solar System and its planets:

The Sun: The Sun is the central star of the Solar System. It is a massive ball of hydrogen and helium, undergoing nuclear fusion to produce light and heat, which sustain life on Earth and drive the dynamics of the Solar System.

A. **Inner Planets (Terrestrial planets)**: Mercury, Venus, Earth, and Mars are closer to the Sun.

Mercury

- Position: Closest planet to the Sun.
- Size: Smallest planet in the Solar System.
- Characteristics: Has a rocky surface with extreme temperature variations; no atmosphere, and no moons.

Venus

- Position: Second planet from the Sun.
- Size: Similar in size and structure to Earth.

- Characteristics: Thick atmosphere of carbon dioxide with clouds of sulfuric acid; hottest planet due to the greenhouse effect; no moons.

Earth

- Position: Third planet from the Sun.
- Size: Largest of the terrestrial (rocky) planets.
- Characteristics**:** Only planet known to support life; has a diverse climate and atmosphere; one moon (the Moon).

Mars

- Position: Fourth planet from the Sun.
- Size: Known as the Red Planet due to iron oxide (rust) on its surface.
- Characteristics: Thin atmosphere mostly of carbon dioxide; has the tallest volcano and the deepest canyon in the Solar System; two small moons (Phobos and Deimos).

B. **Outer Planets (Gas giants and ice giants)**: Jupiter, Saturn, Uranus, and Neptune, shown farther from the Sun.

Jupiter

- Position: Fifth planet from the Sun
- Size: Largest planet in the Solar System
- Characteristics**:** A gas giant with a thick atmosphere of hydrogen and helium; has a Great Red Spot, a massive storm; at least 95 moons, including the largest moon in the Solar System, Ganymede.

Saturn

- Position: Sixth planet from the Sun
- Size: Second-largest planet
- Characteristics: Known for its extensive ring system composed of ice and rock; a gas giant similar to Jupiter; at least 146 moons, with Titan being the largest.

Uranus

- Position: Seventh planet from the Sun
- Size: Third-largest planet
- Characteristics: An ice giant with a blue-green color due to methane in the atmosphere; rotates on its side, which is unique among the planets; at least 28 moons.

Neptune

- Position: Eighth and farthest known planet from the Sun
- Size: Fourth-largest planet
- Characteristics: An ice giant similar to Uranus; has a deep blue color and the strongest winds in the Solar System; at least 16 moons, with Triton being the largest.

Dwarf Planets

- Examples: Pluto, Eris, Haumea, Makemake, and Ceres.

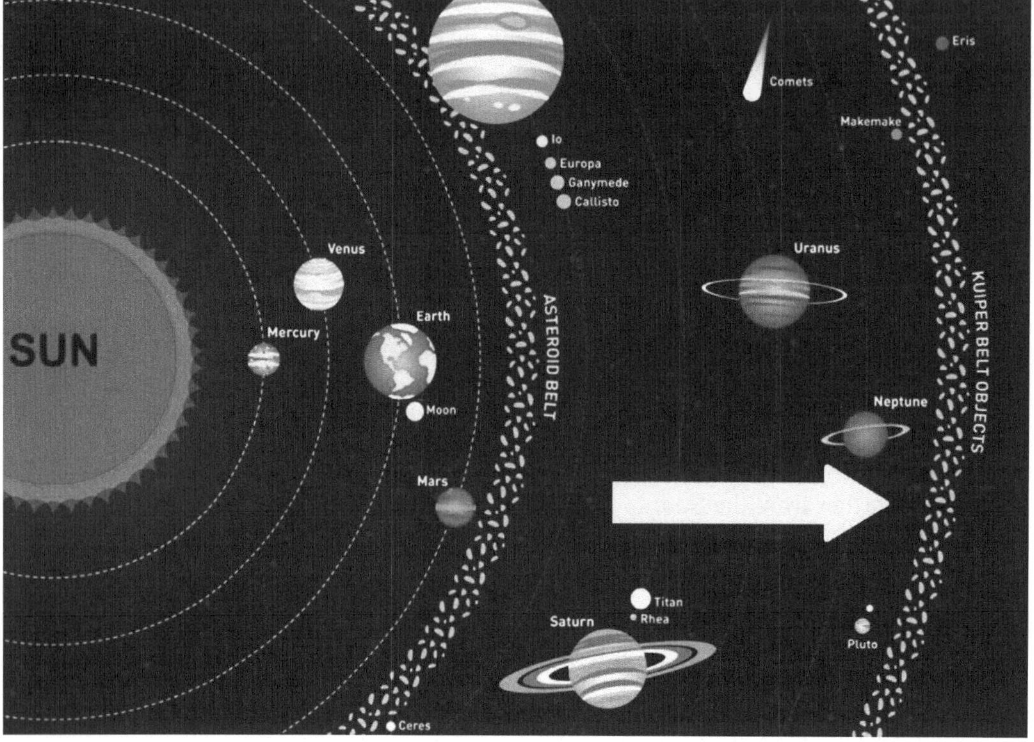

Figure 6.1: Solar system

- Characteristics: Smaller than the eight major planets and not dominant in their orbits; Pluto, once considered the ninth planet, is the most famous dwarf planet.

Other Objects

- Asteroids: Mostly found in the asteroid belt between Mars and Jupiter.
- Comets: Composed of ice, dust, and rock, with long tails that develop as they approach the Sun.
- **Kuiper Belt and Oort cloud:** Regions beyond Neptune filled with icy bodies and potential comets.

The Solar System is vast and complex, with each planet having unique features and characteristics.

Chapter 7

Earth in the Solar System: origin, size, shape, mass, density, rotational and revolution parameters

1. **Origin of Earth:** The Earth was created roughly 4.54 billion years ago as a result of the accumulation of dust and gas inside the protoplanetary disk that surrounded the newly formed Sun.The components that made up this disk were elements that had been produced by stars of earlier times and had been dispersed around the galaxy by supernovae.

Process:

a. **Accretion:** Small particles of dust and ice that were contained inside the disk collided with one another and became entangled, resulting in the formation of bigger entities known as planetesimals.
b. **Proto-Earth:** These planetesimals proceeded to crash with one another and merge together, finally resulting in the formation of a proto-Earth that was molten.
c. **Differentiation:** The Earth went through a process of differentiation as it increased in size. This was a process in which heavier minerals, such as iron, sank to the center, establishing the core, while lighter materials, such as silicates, produced the mantle and the crust.

Moon Formation:

- **Giant Impact Hypothesis:** The Moon of Earth is said to have came into existence not long after the Earth itself was created. Currently, the most prominent explanation proposes that a planet of the size of Mars, known as Theia, collided with the early Earth, resulting in the release of material that later merged together to create the Moon.

2. **Size of Earth**

- Equatorial radius of Earth: 6378.137 km.
- Equatorial diameterof Earth: 12,756.274 km.
- Polar radius of Earth: 6356.752 km.
- Polar diameter of Earth: 12,713.504 km
- Mean radius of Earth: 6371 km.
- Mean diameterof Earth: 12,742 km.
- Equatorial Circumference of Earth: About 40,075 km.
- Polar Circumference of Earth: About 40,008 km.

3. **Shape of Earth**

A. **Oblate Spheroid:** Earth is not a perfect sphere but an oblate spheroid. This means it is slightly flattened at the poles and bulging at the equator due to its rotation. The equatorial bulge causes the equatorial diameter to be about 43 kilometers (27 miles) larger than the polar diameter.

B. **Geoid:** When irregularities of the Earth's surface, such as mountains and ocean trenches, are taken into consideration, the shape of the Earth is more appropriately defined as a geoid. A geoid is a complicated form that is roughly uneven and reflects the mean sea level around the world.

4. **Mass of Earth**

A. **Total Mass:** Earth's mass is approximately 5.972×10^{24} kilograms (1.316×10^{25} pounds).

B. **Gravitational Influence:** The mass of the Earth provides it with a considerable gravitational pull, which is responsible for regulating the Moon's orbit and exerting an impact on other bodies in the Solar System.

5. **Density of Earth:** Earth's average density is 5.513 grams per cubic centimeter (g/cm^3). The Earth's iron core and silicate mantle contribute to its density, making it the planet with the highest density in the Solar System.

6. **Rotational Parameters of Earth**

A. **Rotation Period (Day):** Earth rotates on its axis once every 23.93 hours. This period is known as a sidereal day. However, a solar day (the time it takes for the Sun to return to the same position in the sky) is 24 hours due to the Earth's simultaneous orbit around the Sun.

B. **Axial Tilt:** The axis of the Earth is tilted around its orbital plane, which is known as the ecliptic plane, at an angle of 23.5 degrees. Because of this axial tilt, various regions of the Earth get different amounts of sunlight throughout the year. This is what causes the seasons to occur.

C. **Rotational Velocity:** The Earth rotates at a speed of around 1,674 kilometers per hour (1,040 miles per hour) at the equator. As one approaches the poles, the rotational velocity decreases until it is almost zero.

7. **Revolution Parameters of Earth**

A. **Orbital Period (Year):** According to estimates, it takes roughly 365.25 days for the Earth to complete one circle around the Sun. The time period in consideration is referred to as a sidereal year. An additional leap day is added every four years, which is known as a leap year, in order to adjust for the additional 0.25 days.

B. **Orbital Shape:**Since the Sun is located at one of the foci of the ellipse that constitutes the orbit of the Earth, this means that the orbit is not a complete circle but rather slightly elongated. An astronomical unit (AU) is a unit of measurement that is used to describe the distance between the Earth and the Sun. This distance is about 149.6 million kilometers (93 million miles).

C. **Orbital Velocity:** It is estimated that the average orbital velocity of the Earth around the Sun is around 29.78 kilometers per second, which is equivalent to 107,000 kilometers per hour or 67,000 miles per hour.

8. **Milankovitch Cycles:** The Milankovitch cycle describes how changes in the tilt and orbit of the Earth impact the climate over extended periods of time, creating glacial and interglacial periods. These cycles, which are named after the Serbian astronomer and mathematician Milutin Milankovitch, are made up of three primary elements that influence Earth's climate via orbital changes:

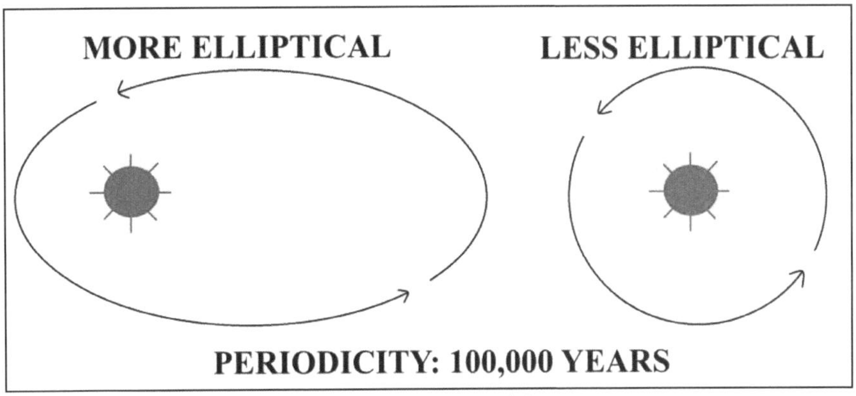

Figure 7.1: Eccentricity

A. **Eccentricity (~100,000 years or 100 kyr):** Over a period of around 100,000 years, the Earth's orbit around the Sun changes form, becoming more elliptical and less circular. The amount of solar energy that the Earth gets is altered as a result of this, particularly at certain periods of the year. This is because the distance between the Earth and the Sun is altered.

B. **Axial Tilt (Obliquity) (~41,000 years or 41 kyr):** Over the course of roughly 41,000 years, the angle at which the Earth's axis is oriented undergoes a minor shift, fluctuating between around 22.1 and 24.5 degrees. The tilt has an effect on the intensity of the seasons, with larger tilts resulting in more extreme seasonal changes and lesser tilts resulting in seasons that are more moderate.

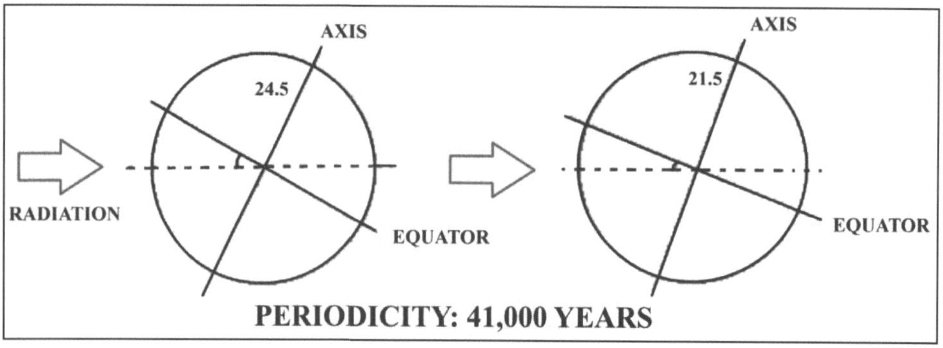

Figure 7.2: Obliquity

C. **Precession (~22,000 years or 22 kyr):** In a cycle that lasts around 22,000 years, the axis of the Earth will shift its orientation, causing it to wobble throughout the course of time. The timing of the seasons shifts as a result of this alteration in respect to the location of the Earth in its orbit, which may modify the patterns of climate.

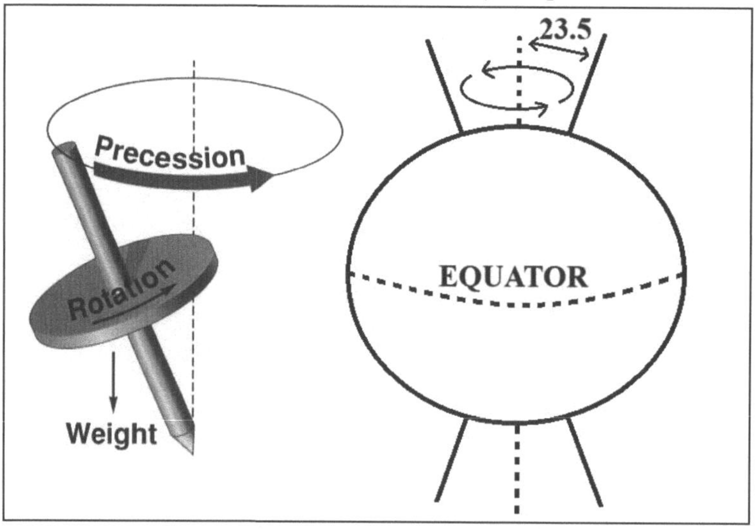

Figure 7.3: Diagram illustration the Precession of earth

These three orbital variations combine to affect the distribution and intensity of sunlight received by Earth, driving long-term climate changes such as ice ages.

9. Summary

- Earth is a dynamic planet with a rich history of formation that has resulted in a unique set of characteristics.
- It is the third planet from the Sun, with a diameter of about 12,742 kilometers, and a mass of 5.972×10^{24} kilograms.
- The planet's shape is an oblate spheroid, and it has an average density of 5.513 g/cm^3.
- Earth rotates on its axis every 23.93 hours and revolves around the Sun in about 365.25 days.
- These physical parameters and movements not only define Earth's day and year but also its ability to support life, a feature that sets it apart in the Solar System.

Chapter 8
The Concept of Geoid and Spheroid

Introduction

One of the most important aspects of geodesy, cartography, and navigation is having a solid understanding of the shape of the Earth. The geoid and the spheroid (or ellipsoids) are two notions that are essential when attempting to describe the shape of the Earth. Having these models allows for more accurate measurements of places on the surface of the Earth, as well as greater understanding of the geometry and gravitational field of the Earth.

The Geoid: The geoid is an equipotential surface of the Earth's gravity field which coincides with the mean sea level in the oceans and extends under the continents. It represents the true physical shape of the Earth, reflecting variations in the Earth's gravitational field.

A. Characteristics

a. Equipotential Surface: When the gravitational potential is constant on a surface, this is referred to as the geoid. The result of this is that a tiny mass would not experience any net force if it were to be transported over this surface.

b. Irregular Shape: The geoid is not a simply geometric form; rather, it is irregular because of differences in the density of the Earth and the distribution of mass.

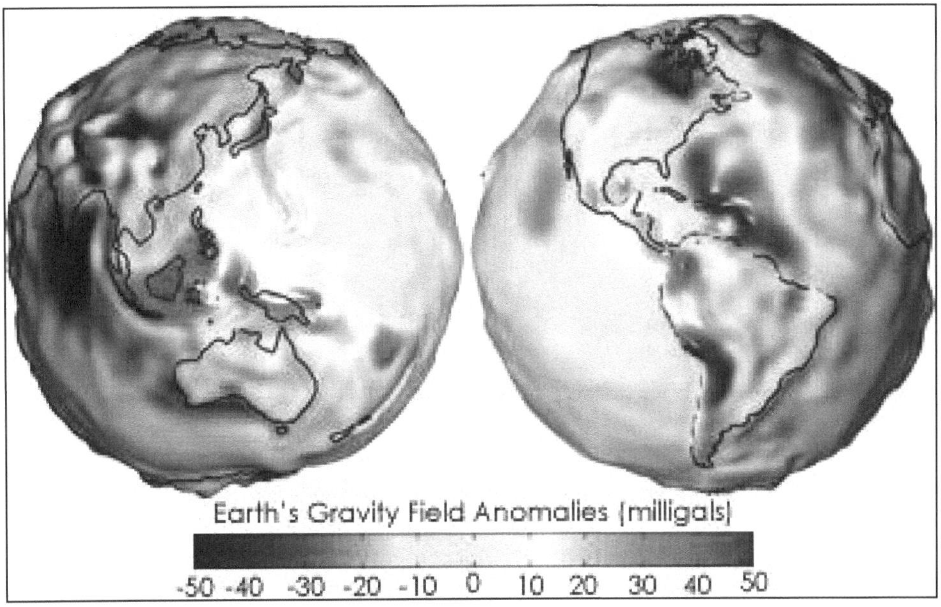

Figure 8.1: Earth Geoid structure (Reference: GRACE: Data Processing and Products." *Earth Observatory*, NASA)

B. Determination of the Geoid

a. **Gravity Measurements:** By using gravimeters and satellite measurements, the geoid may be calculated by measuring the gravitational field of the Earth.

b. **Sea Level Data:** Satellite altimetry and tide gauges are two methods that are used to get measurements of sea level, which are then utilized to improve the geoid model.

C. Applications

a. **Height Measurement:** In the process of measuring elevations, the geoid is used as a reference. Orthometric heights are heights that are measured above the geoid.

b. **Satellite Navigation:** The accuracy of the Global Positioning System (GPS) and other satellite-based positioning systems is improved by accurate geoid models.

c. **Geophysical Studies:** The study of geological processes is aided by the information that is provided by variations in the geoid, which provides information on the distribution of mass inside the Earth.

The Spheroid (Ellipsoid): A spheroid (or ellipsoid) is a mathematically defined surface that approximates the shape of the Earth. It is formed by rotating an ellipse about its minor axis (oblate spheroid) or major axis (prolate spheroid).

A. Characteristics

a. **Oblate Spheroid**: It is possible to make the most accurate representation of the Earth by using an oblate spheroid, in which the equatorial radius is greater than the polar radius because of the rotation of the Earth.

b. **Mathematical Model:** Two parameters—the equatorial radius and the polar radius—are used to define the spheroid, which is a simplified geometric form.

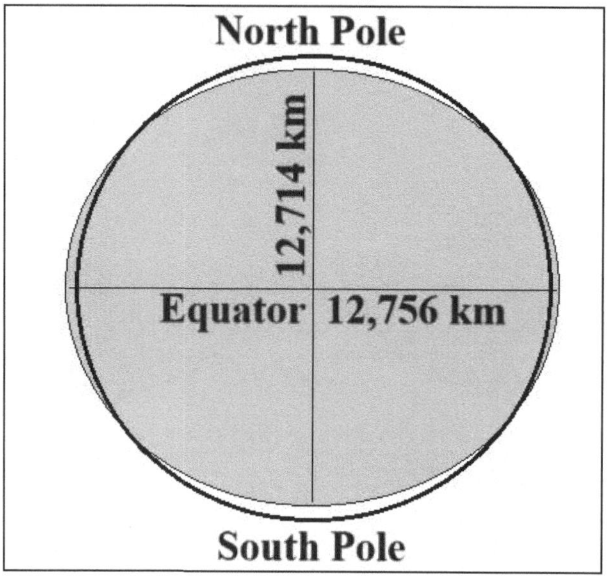

Figure 8.2: Earth Spheroid (Ellipsoid)

B. **Common Spheroid Models**

a. **WGS84 (World Geodetic System 1984)**: Equatorial Radius: 6,378.137 km, and flattening: 1/298.257223563. Widely used in GPS and global mapping applications.

b. **GRS80 (Geodetic Reference System 1980)**: Equatorial Radius: 6,378.137 km and flattening: 1/298.257222101. Commonly used for geodetic calculations and mapping.

C. **Applications**

a. **Geodesy and Cartography:** For the purposes of mapping and geodetic calculations, spheroids serve as reference surfaces, which make it possible to provide a consistent framework for coordinate systems.

b. **Satellite Orbits:** The mathematical calculation of satellite orbits and locations is impacted by the shape of the spheroid.

c. **Navigation:** Models of spheroid are very necessary for the development of accurate navigation and positioning systems.

Comparison between Geoid and Spheroid

A. **Conceptual Differences**

a. **Nature:**

- Geoid: A physical, irregular surface based on the Earth's gravity field.
- Spheroid: A mathematical, regular surface approximating the Earth's shape.

b. **Reference:**

- Geoid: Reflects mean sea level and gravity variations.

- Spheroid: Defined by geometric parameters (radius and flattening).

B. Practical Differences

a. Precision:

- Geoid: Provides a more accurate representation of the Earth's physical shape, essential for precise height measurements.
- Spheroid: Easier to use for calculations and mapping, but less precise in representing the Earth's true shape.

b. Use Cases:

- Geoid: Used in geophysics, oceanography, and height determination.
- Spheroid: Used in geodesy, cartography, and satellite navigation.

Integration of Geoid and Spheroid

A. **Geoid Undulations:** Undulations on the geoid are the physical characteristics that differentiate it from the spheroid. These undulations are applied to convert between heights measured corresponding to the geoid (orthometric heights) and heights measured compared to the spheroid (ellipsoidal heights). These undulations vary on a global scale.

B. Geoid Models

a. Global Geoid Models:

- **EGM2008 (Earth Gravitational Model 2008):** Provides a high-resolution global geoid model.
- **GOCE (Gravity Field and Steady-State Ocean Circulation Explorer):** Satellite mission that provided detailed data for refining the geoid model.

b. Local Geoid Models: Countries and regions develop their own geoid models to improve the accuracy of local height measurements and geodetic applications.

C. Practical Use in Height Measurement

a. Height Systems:

- **Orthometric Height:** Measured from the geoid, commonly used in engineering and topographic maps.
- **Ellipsoidal Height:** Measured from the spheroid, used in GPS systems.

b. Conversion: Geoid models are used to convert between Orthometric and ellipsoidal heights, ensuring consistency across different height measurement systems.

A basic knowledge of the form of the Earth and its gravitational field requires an understanding of the geoid and the spheroid concepts. Unlike the spheroid, which is a simplified mathematical model used for geodetic calculations and mapping, the geoid is an illustration of the actual physical form of the Earth, which is a reflection of the gravitational changes. In the areas of navigation, satellite positioning, and geophysical research, the integration of both models permits accurate height measurements and provides more precise measurement.

Chapter 9
Kepler's Laws of Planetary Motion

Around the beginning of the 17th century, the German astronomer and mathematician Johannes Kepler developed three basic rules of planetary motion. These rules that define the motion of planets around the Sun are fundamental to understanding the dynamics of the Solar System because they explain how the planets move around the Sun.

1. **Kepler's First Law: The Law of Ellipses**

A. **Statement**: "The orbit of every planet is an ellipse with the Sun at one of the two foci."

B. **Explanation**: Planets do not orbit the Sun in perfect circles but rather on elliptical orbits, which are represented by the form of an oval in shape. A closed curve with two focus points is called an ellipse. The Sun is not at the center of the ellipse; rather, it is in one of these foci. Throughout its orbit, a planet's distance from the Sun changes. The planet is closest to the Sun during its perihelion, or closest point. The planet is at its aphelion, or most far point from the Sun.

C. **Significance**: The idea that planetary orbits were perfect circles, which had been maintained since ancient Greek astronomy, has been disproved by this rule. It paved the way for the realization that certain rules, rather than arbitrary paths, control celestial movements.

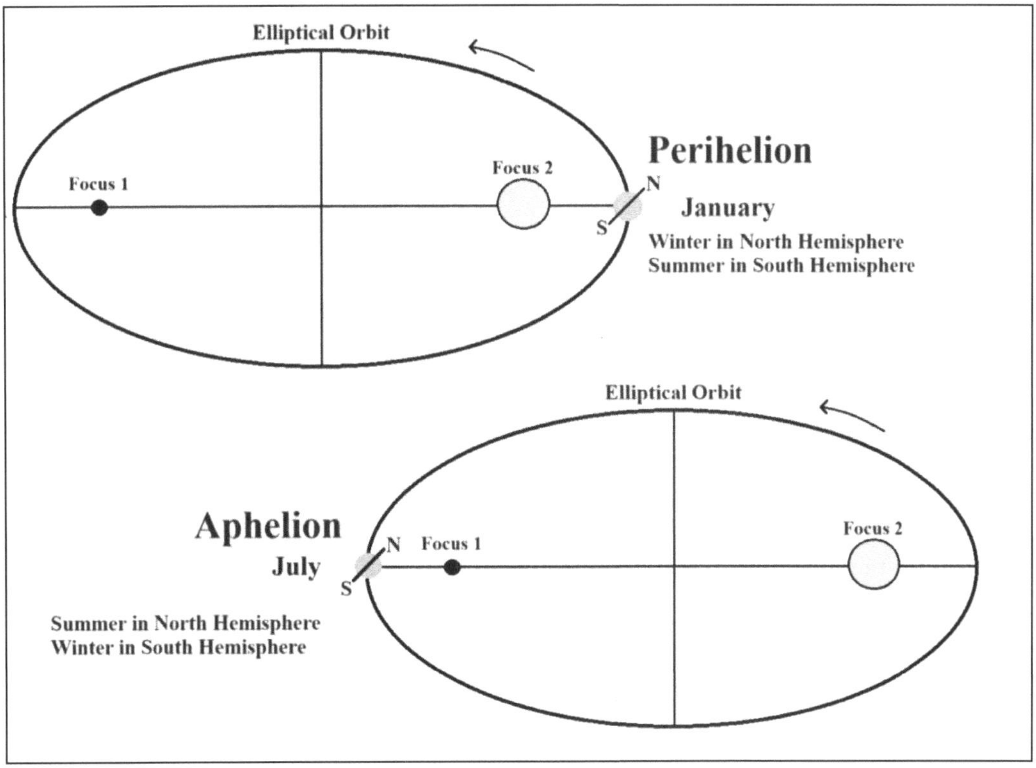

Figure 9.1- Showing kepler's first law

D. **Real-World Example**: The Earth's orbit around the Sun is elliptical; in early January, it is at perihelion, or the closest point to the Sun, and in early July, it is at aphelion, or the farthest point from the Sun (Figure 9.1).

2. Kepler's Second Law: The Law of Equal Areas

A. **Statement**: "A line segment joining a planet and the Sun sweeps out equal areas during equal intervals of time."

B. **Explanation**: The imaginary line that connects a planet to the Sun, called the radius vector, sweeps out equal portions at equal periods of time as the planet revolves around the Sun. Accordingly, a planet travels more slowly during aphelion, when it is further from the Sun, and more quickly at perihelion, when it is closer to the Sun (Figure 9.2).

C. **Implications**:

a. **Variable Speed**: Planets travel slowly when they are distant from the Sun and more quickly when they are closer to it. This variation in speed is necessary to ensure that the area swept out by the radius vector remains constant over equal time intervals.

b. **Conservation of Angular Momentum**: Kepler's Second Law is a direct consequence of the conservation of angular momentum. In order to maintain its angular momentum, the planet had to accelerate its motion as it came closer to the sun.

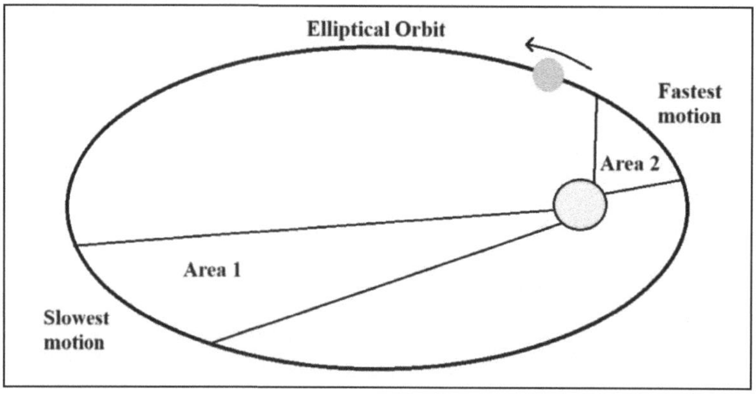

Figure 9.2- Showing kepler's Second law

D. **Significance**: This principle explains how a planet's orbital speed changes over time, indicating that planetary motion is not constant but rather fluctuates according to the planet's distance from the Sun. It results from the conservation of angular momentum, which states that a planet's velocity must adapt in order to maintain energy conservation throughout its elliptical orbit.

E. **Real-World Example**: Think about how Mars orbits the Sun. When Mars is nearer the Sun, it moves more quickly and covers more distance in the same period of time as when it is further away (Figure 9.2).

3. Kepler's Third Law: The Law of Harmonies

A. **Statement**: "The square of the orbital period of a planet is directly proportional to the cube of the semi-major axis of its orbit." According to Kepler's Third Law, the amount of time it takes for a planet to complete one orbit around the sun is proportional to the radius of the planet's orbit.

B. **Mathematical Form**:

$$T^2 \propto \alpha^3$$

Where

- T is the orbital period (the time it takes for a planet to complete one orbit around the Sun)
- α is the semi-major axis (the longest radius of the ellipse).
- The proportionally can be written as:

$$\frac{T^2}{a^3} = \textbf{constant}$$

- For the planets in the solar system, this constant is the same.

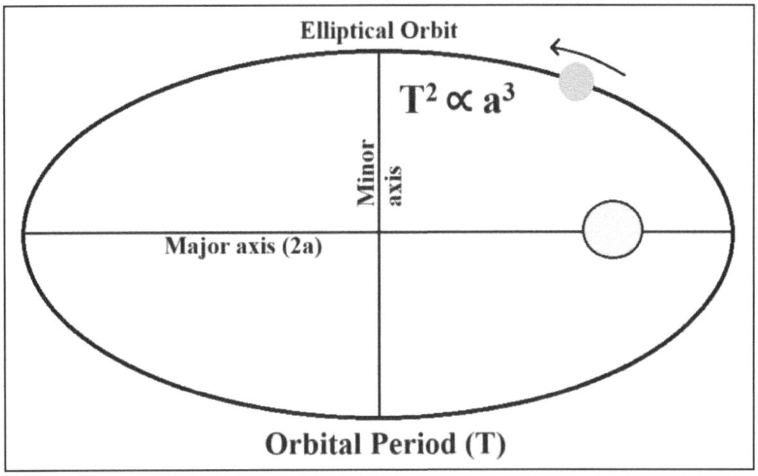

Figure 9.3 -Showing kepler's third law

C. **Implications**:

a. **Relationship between Distance and Orbital Period**: There is a correlation between planets that are positioned more distant from the Sun and having orbital periods that are longer. For example, since Saturn is situated at a larger distance from the Sun than Earth is, it takes a substantially longer period of time for it to complete one orbit around the Sun.

b. **Universal Application**: Although Kepler developed this equation for the Solar System's planets, Newton subsequently demonstrated that it applies universally to any two bodies circling one another while under the influence of gravity.

A. **Explanation**: This law establishes a precise mathematical relationship between the time it takes a planet to orbit the Sun and the size of its orbit. For any planet in the Solar System, if you take the square of its orbital period (in Earth years) and divide it by the cube of the semi-major axis of its orbit (in astronomical units, AU), the result is the same for all planets.

B. **Significance**: The third law provided a tool to compare the motions of different planets and to predict their positions at any given time. It implies that planets farther from the Sun take much longer to complete an orbit than those closer to the Sun. This law is universal, applying not just to planets, but to any object in a stable orbit around a central body.

4. Significance of Kepler's Laws

A. **Shift from Geocentric to Heliocentric Model**: Kepler's laws provided strong evidence for the heliocentric model, where planets orbit the Sun, rather than the Earth-centered model that had been dominant for centuries.

B. **Foundation for Newtonian Mechanics**: Kepler's laws were crucial in the development of Newton's law of universal gravitation. Newton was able to derive Kepler's laws from his gravitational theory, demonstrating that the same physical laws apply both on Earth and in the heavens.

C. **Predictive Power**: These laws allowed astronomers to predict planetary positions with great accuracy, facilitating the advancement of celestial mechanics and navigation.

D. **Broader Application**: Kepler's laws are not limited to the Solar System; they apply to any system where two bodies are in gravitational orbit, including exoplanets, binary stars, and even galaxies orbiting each other.

5. Historical Context and Development

A. **Tycho Brahe's Observations**: Kepler worked as an assistant to Tycho Brahe, who had compiled the most accurate astronomical data of the time. Kepler used this data, particularly the observations of Mars, to formulate his laws.

B. **Elliptical Orbits**: Kepler's realization that planetary orbits are elliptical was revolutionary, as it contradicted the long-held belief in circular orbits that had dominated since ancient Greek astronomy.

C. **Publication**: Kepler published his first two laws in 1609 in his work *Astronomia Nova*. The third law was published later in 1619 in his book *Harmonices Mundi* (The Harmony of the World).

Kepler's laws of planetary motion revolutionized our understanding of the Solar System, providing a clear, mathematical description of how planets move around the Sun. These laws continue to be fundamental in the fields of astronomy, physics, and space exploration.

Chapter 10
The Chemical Composition of the Earth

The Earth is composed of various elements, minerals, and compounds that are distributed unevenly between its different layers: the crust, mantle, outer core, and inner core. Understanding the chemical composition of the Earth provides insight into its formation, structure, and the dynamic processes that occur within.

1. Overview of Earth's Chemical Composition

A. Elemental Composition:

a. The Earth is primarily composed of a few key elements, with oxygen, silicon, aluminum, iron, calcium, sodium, potassium, and magnesium being the most abundant.

b. These elements combine to form various minerals and rocks that make up the Earth's crust, mantle, and core.

B. Layered Structure: The Earth is mainly divided into three layers, each with distinct chemical compositions:

a. Crust: The outermost layer composed mostly of silicate minerals.

b. Mantle: A thick layer of silicate rock that lies beneath the crust.

c. Core: Divided into the outer core (liquid) and inner core (solid), primarily composed of iron and nickel.

2. **Chemical Composition of the Earth's Crust:** The crust is the Earth's outermost layer, varying in thickness from about 5-70 kilometers. The Earth's crust is about 95% igneous rocks by volume, with the remaining 5% being sedimentary and metamorphic rocks. It includes both continental and oceanic crust.

A. **Continental Crust**:

a. Thickness: 30-70 kilometers on average.

b. Composition: Primarily composed of granitic rocks, which are rich in silica and aluminum.

c. **Major Elements**:

1. Oxygen (O): ~46.6% by weight.
2. Silicon (Si): ~27.7% by weight.
3. Aluminum (Al): ~8.1% by weight.
4. Iron (Fe): ~5.0% by weight.
5. Calcium (Ca): ~3.6% by weight.
6. Sodium (Na): ~2.8% by weight.
7. Potassium (K): ~2.6% by weight.
8. Magnesium (Mg): ~2.1% by weight.

Minerals: The most common minerals in the continental crust are silicates, including feldspar (plagioclase and orthoclase), quartz, mica, and amphiboles.

B. **Oceanic Crust**:

a. Thickness: 5-10 kilometers on average.

b. Composition: Primarily composed of basaltic rocks, which are rich in iron and magnesium.

c. Major Elements: Similar to the continental crust but with higher proportions of iron and magnesium and lower proportions of silica and aluminum.

d. Minerals: Dominated by plagioclase feldspar, pyroxenes, and olivine.

3. **Chemical Composition of the Earth's Mantle:** The mantle lies between the crust and the core, extending from a depth of about 35 kilometers to 2,900 kilometers.

a. **Major Elements**:

1. Oxygen (O): ~44.8% by weight.
2. Magnesium (Mg): ~22.8% by weight.
3. Silicon (Si): ~21.5% by weight.
4. Iron (Fe): ~6.3% by weight.
5. Calcium (Ca): ~2.5% by weight.
6. Aluminum (Al): ~2.2% by weight.

b. **Minerals**:

- The mantle is composed predominantly of silicate minerals rich in magnesium and iron, such as olivine $(Mg,Fe)_2SiO_4$, pyroxenes $(Mg,Fe)SiO_3$, and garnet.

- **Peridotite:** A rock made mostly of olivine and pyroxenes, is the dominant rock type in the upper mantle.
- The mantle also contains minerals like spinel, perovskite, and wadsleyite at greater depths.

Divisions of the Mantle:

A. Upper Mantle: Extends from the base of the crust to about 660 kilometers depth; it includes the lithosphere (rigid) and asthenosphere (plastic, partially molten).

B. Transition Zone: Between 410 and 660 kilometers depth, where high-pressure minerals like wadsleyite and ringwoodite are stable.

C. Lower Mantle: Extends from 660 kilometers to 2,900 kilometers, composed of high-pressure silicate minerals like bridgmanite and ferropericlase.

4. Chemical Composition of the Earth's Core: The core is divided into the outer core (liquid) and inner core (solid), extending from about 2,900 kilometers to the center of the Earth at about 6,371 kilometers.

A. Outer Core:

a. Composition: Primarily composed of iron (Fe) and nickel (Ni), with lighter elements such as sulfur (S), oxygen (O), and possibly silicon (Si) in smaller amounts.

b. State: Liquid, due to the high temperatures that prevent the iron-nickel alloy from solidifying.

B. Inner Core:

a. Composition: Similar to the outer core, primarily iron (Fe) with some nickel (Ni) and trace amounts of lighter elements.

b. State: Solid, due to the extreme pressure that forces the iron to crystallize despite the high temperature.

c. Elemental Distribution:

- Iron (Fe): ~85% by weight in the core.
- Nickel (Ni): ~5-10% by weight.
- Lighter Elements: ~5-10% by weight, with candidates including sulfur, silicon, and oxygen.

d. Significance: The Earth's magnetic field is generated by the motion of the liquid outer core, which involves the flow of molten iron and nickel, creating a dynamo effect.

5. Minor and Trace Elements in the Earth

A. Trace Elements:

a. The Earth also contains trace amounts of other elements, such as carbon (C), hydrogen (H), phosphorus (P), and noble gases (helium, neon, argon).

b. These elements are critical for life and are found in small concentrations throughout the Earth's crust and mantle.

B. **Rare Earth Elements**: The Earth's crust contains small amounts of rare earth elements (REEs), such as lanthanum, cerium, and neodymium, which are important for modern technologies like electronics and renewable energy systems.

6. **Evolution of Earth's Chemical Composition**

A. Formation: The Earth formed about 4.54 billion years ago from the solar nebula, a cloud of gas and dust. The accretion of material led to the differentiation of the Earth into layers based on density and composition.

B. Differentiation: Early in Earth's history, it underwent differentiation, where heavier elements like iron and nickel sank to form the core, while lighter silicate materials formed the mantle and crust.

C. Geological Processes: Plate tectonics, volcanic activity, erosion, and other geological processes have continually reshaped the Earth's surface, redistributing elements and altering the chemical composition of different layers.

D. Continual Change: The Earth's chemical composition is not static; it continues to evolve due to ongoing geological processes, including the cycling of elements between the Earth's interior, surface, and atmosphere.

The chemical composition of the Earth is complex and varies significantly between the different layers of the crust, mantle, and core. Oxygen, silicon, aluminum, and iron are the most abundant elements in the crust, while the mantle is rich in magnesium and silicate minerals, and the core is primarily composed of iron and nickel. This layered structure and the distribution of elements are key to understanding the Earth's physical properties, its magnetic field, and the dynamic processes that shape our planet.

Chapter 11
Basic Concepts of Seismology and the Internal Structure of the Earth

The scientific investigation of earthquakes and the transmission of elastic waves through the Earth or other planet-like bodies are referred to as seismology. It plays a vital significance to acquire knowledge of the interior structure of the Earth, which may be divided down into three distinct layers: the crust, the mantle, and the core.

Basic Concepts of Seismology:

The field of seismology is concerned with the study of seismic waves, which may be produced by catastrophic events such as earthquakes or by man-made causes such as explosions. Information about the interior of the Earth may be obtained from these waves.

Earthquakes: An earthquake occurs when there is a rapid release of energy in the Earth's lithosphere as a result of volcanic eruptions, tectonic activity, or human activities like mining.

Focus (Hypocenter): The location of the origin of the earthquake is referred to as the focus, which is also known as the hypocenter.

Epicenter: The location on the surface of the Earth that is directly above the focus is known as the epicenter.

Seismic Waves: The energy waves that flow through the Earth as a consequence of an earthquake, explosion, or sudden mechanical impulse are referred to as seismic waves.

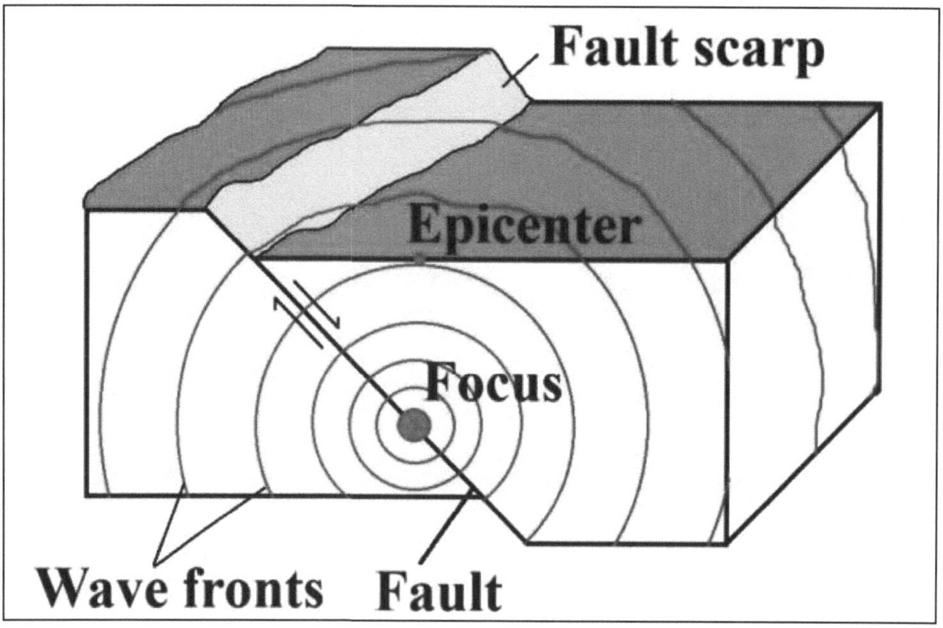

Figure 11.1: Basic terminology of associated with Earthquakes

Types of Seismic Waves:

Body waves and surface waves are the two primary types of waves that are produced by earthquakes. There are various characteristics and behaviors associated with each category.

A. **Body Waves**: Body waves are the first to be observed in seismographs, since they travel into the interior of the Earth. Primary (P) waves and secondary (S) waves are the two kinds of waves that are further subdivided into these categories.

1. **P-Waves (Primary or Compressional Waves)**:

A. **Speed:** P waves are the kind of seismic waves that pass through solids, liquids, and gases at the fastest travelling speed. They travel at rates ranging from 5 to 8 kilometers per second under the crust of the Earth.

B. **Motion:** They are compressional waves that move by pushing and pulling the ground in the direction of wave propagation (similar to sound waves). These waves travel through and out of the earth.

C. **Characteristics:** P-waves are the first to be detected by seismometers after an earthquake. It is usually true that P-waves impose less damage than other types of waves since they do not have the strong side-to-side or up-and-down motion that may cause damage to materials and buildings.

2. **S-Waves (Secondary or Shear Waves)**:

A. **Speed:**The speed of S waves in the Earth's crust ranges from 3 to 4 km/s, which is slower than the speed of P waves. Solids are the only medium through which they can travel.

B. **Motion:** Perpendicular to the direction of wave propagation, causing a shearing motion.

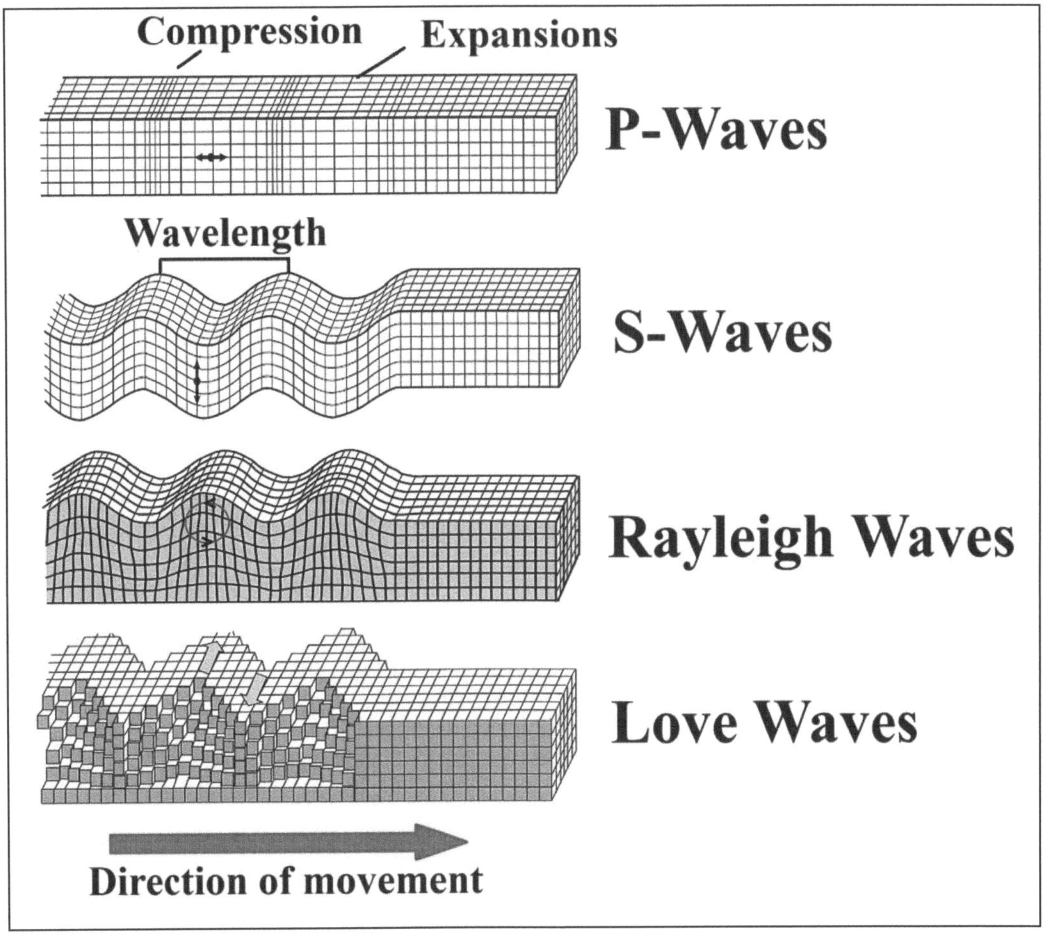

Figure 11.2: Types of Seismic Waves

C. **Characteristics:** Due to their more powerful movement, S-waves are responsible for more damage than P-waves; yet, they still are normally responsible for less damage than surface waves. Considering that S-waves come after P-waves and are unable to pass through the liquid outer core, this information helps develop a better understanding of the interior structure of the Earth.

B. Surface Waves: Surface waves are waves that travel along the surface of the earth. Although surface waves move at a slower speed than body waves, they are capable of causing far more damage because of to their large amplitude and the way in which they travel over the surface of the Earth. There are two primary kinds of these waves, which are known as love waves and Rayleigh waves.

1. Rayleigh Waves:

A. Motion: Rayleigh waves travel in an elliptical motion, which is equivalent to the motion of ocean waves, and they produce movement of the ground in both the vertical and horizontal directions.

B. Speed: Slower than both P-waves and S-waves.

C. Characteristics: They also contribute significantly to surface destruction but are typically slower than Love waves

2. Love Waves:

A. Motion: Motion perpendicular to the direction of propagation cause of horizontal shearing of the ground.

B. Speed: Generally faster than Rayleigh waves but slower than S-waves.

C. Characteristics: There is a kind of surface wave known as love waves, which are the most rapid of all surface waves. In a horizontal motion that is perpendicular to the direction of propagation, they move the ground from side to side. These side-to-side motions near the surface can cause serious damage to buildings, infrastructure, and the surface of the Earth, which is why they are responsible for the most damage that occurs during an earthquake.

Why L-waves Cause the Most Damage:

A. Amplitude: Body waves have lower amplitudes than surface waves, which mean that surface waves, especially L-waves, generate more ground displacement than body waves.

B. Location: As a consequence of the fact that they travel over the surface of the Earth, the shaking is more noticeable near the surface, more specifically in areas where buildings and structures are built.

C. Horizontal motion: Through their horizontal motion, Love waves have the potential to force structures to violently sway, which often results in significant structural damage, especially to the foundations and walls of the building.

Seismographs and Seismogram:

Seismographs are instruments that are used to detect, record, and measure vibrations that are generated by seismic waves. These vibrations are often created by earthquakes or other seismic occurrences. It consists of a mass suspended from a fixed base; when the ground shakes due to earthquakes the mass tends to stay still due to inertia, while the base moves along with the ground. The instrument plays a part for recording the relative motion that occurs between the mass and the base. The actual record or graph that is created by the seismograph is referred to as a seismogram. The seismic waves are shown as a sequence of lines that change in amplitude and frequency. These lines represent the strength of the ground motion as well as the duration of the ground motion. By analyzing these patterns, experts are able to determine the magnitude of the seismic event, as well as its location and characteristics.

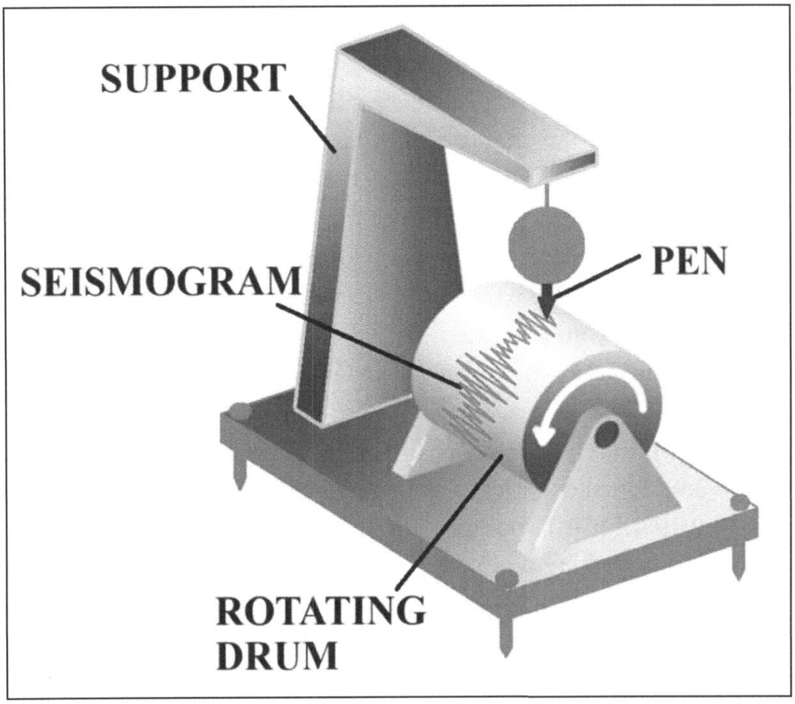

Figure 11.3: Seismographs and seismogram

Measurement of Earthquakes

Richter scale – Earthquake Magnitude:

Charles F. Richter, an American seismologist, developed the Richter scale in 1935. It is a logarithmic scale that is used to estimate the magnitude of earthquakes. For the purpose of determining the magnitude of an earthquake, it measures the amplitude of seismic waves that have been recorded by seismograph instrument.

How It Works:

A. **Logarithmic Nature:** The scale is logarithmic, which means that an increase of one full number equates to a tenfold increase in the amplitude that is being measured and about 31.6 times more energy being released. For example, a magnitude 5.0 earthquake releases about 31.6 times more energy than a magnitude 4.0 earthquake.

B. **Measurement:**Both the amplitude of the biggest seismic wave that was recorded and the distance between the epicenter of the earthquake and the seismograph station are used in the calculation of the Richter magnitude.

Limitations:

A. **Magnitude Saturation:**Due to the fact that the Richter scale was first developed for moderate-sized earthquakes in Southern California, it is less accurate for earthquakes that are greater than 7.0 magnitudes.

B. **Regional Specificity:**Due to the fact that it was calibrated based on certain instruments and geological conditions, it is less effective in other places or when used with alternative equipment.

Modern Alternatives: The Richter scale has been mainly replaced by the Moment Magnitude Scale (Mw) for the purpose of measuring earthquakes worldwide that are medium to large in magnitude. This is due to the limitations of the Richter scale. The seismic moment, which takes into account both the region of the fault that slipped and the amount of slip, is taken into account by the Moment Magnitude Scale, which offers a more accurate representation of the magnitude of an earthquake.

Richter scale Magnitude Categories:

1. **Micro (Less than 2.0)**: can't experience or feel by people, detected only by seismographs.
2. **Minor (2.0 to 2.9)**: Rarely experience or feel by people but can be recorded.
3. **Light (3.0 to 3.9)**: Often experience or feel by people, but usually causes no damage.
4. **Moderate (4.0 to 4.9)**: It is possible for structures to overcome slight damage and generate observable shaking within the building.
5. **Strong (5.0 to 5.9)**: Has the potential to inflict damage to structures, particularly those that are poorly built.
6. **Major (6.0 to 6.9)**: Causes significant damage in populated areas.
7. **Great (7.0 to 7.9)**: Can cause serious damage in a large region.
8. **Massive (8.0 and above)**: Causes catastrophic damage can destroy communities over vast areas.

Historical Significance: Magnitudes of earthquakes were first measured using the Richter scale, which was the first standard scale. It made it possible to compare the magnitudes of earthquakes independent of the place or time period in which the event occurred.

Key Takeaways: On the basis of the amplitudes of seismic waves, the Richter scale is used to measure the magnitude of earthquakes. Each step represents a large increase in the amount of energy that is released, and the process is logarithmic. In modern seismology, the Moment Magnitude Scale is often used for the purpose of providing more accurate measurements of major earthquakes.

Intensity: Mercalli Scale – Earthquake Damage

The Mercalli Intensity Scale is a seismic scale that is used for the purpose of determining the intensity of an earthquake by analyzing the impacts that it has with regard to people, buildings, and the surface of the Earth. Compared to the Richter scale, which measures the amount of energy that is produced at the epicenter of the earthquake (its magnitude), the Mercalli Scale analyzes the effect that the earthquake had at specific locations where it occurred.

Historical Background:

In the year 1902, the initial scale was designed by Giuseppe Mercalli, an Italian priest an d volcanologist. Eventually, in 1931, American seismologists Harry O. Wood and Frank Neumann made modifications to the scale, which resulted in the creation of the Modified Mercalli Intensity Scale (MMI), which is widely used today.

Scale Structure

The Modified Mercalli Intensity Scale is comprised of 12 gradually more intense levels, which are denoted by Roman numerals ranging from I to XII. The intensity of the shaking and damage that occurred during an earthquake is represented by each level.

Intensity Levels and Descriptions

1. **I (Not Felt)**: Under extremely conditions, only a small number of people experience earthquakes.
2. **II (Weak)**: Experienced by a small number of persons, particularly in higher floors of buildings.
3. **III (Weak)**: Particularly on higher floors, it feels very much inside the building. Many individuals do not recognize that it is an earthquake.
4. **IV (Light)**: Many people feel inside, whereas few feel outside. A few woke up. Doors and windows vibrate.
5. **V (Moderate)**: Experienced by almost everyone. A lot of people are awakened. Windows and some dishes are broken. Unstable items turned over.
6. **VI (Strong)**: Experienced by all; many people are frightened. Some heavy furniture moves. Minor structural damage.
7. **VII (Very Strong)**: Damage is minor in well-constructed buildings, low to moderate in average buildings, and there is noticeable cracking.
8. **VIII (Severe)**: significant harm to regular structure. Collapse may occur partially. Heavy furniture will fall over.
9. **IX (Violent)**: Significant damage to large structures. Structures may move away from their foundations. Pipes break and the ground fractures.
10. **X (Extreme)**: Most masonry and frame buildings will be destroyed, along with some well-built timber ones. The ground has significant cracks.

11. **XI (Extreme)**: Few, if any, masonry structures remain standing. Bridges destroyed. Landslides occur. Broad fissures in ground.
12. **XII (Catastrophic)**: Nearly total collapse. Vibrations that is visible on the ground. Items thrown into the air.

Key Features

A. **Subjective Assessment:** As it is based on human observations and experiences that have been recorded, the scale is dependent on subjectivity.

B. **Location-Specific Intensity:** Depending on factors such as distance from the epicenter, local geology, and typical construction methods, the intensity of an earthquake may vary greatly from one region to another.

C. **Damage Correlation:** Higher intensity levels correlate with greater damage to structures and infrastructure.

Applications

A. **Engineering and Construction:** Provides help to engineers in the design of structures and infrastructure that are capable of withstanding the expected seismic intensities in an area.

B. **Seismic Hazard Mapping:**Provides assistance in the design of maps that illustrate regions of possible risk based on reports of previous intensity.

C. **Emergency Response:** Supports in the disaster response and resource allocation by providing timely evaluations of the severity of the damage.

Comparison with Magnitude Scales

A. **Mercalli Scale (Intensity):** Subjective and qualitative. Measures observed effects and damage. Varies by location.

B. **Richter/Moment Magnitude Scales (Magnitude):** Objective and quantitative. Measure energy released at the source. Single value for each earthquake.

Limitations

A. **Subjectivity:** Relies on personal reports, which can be influenced by human perception and bias.

B. **Population Density Dependency:** As a consequence of fewer observations, it is less effective in regions with a low population density.

C. **Modern Building Codes:** There is a possibility that same intensity levels may result in less harm as a consequence of advances in engineering, which will influence consistency over time.

Historical Significance: Prior to the invention of contemporary seismographs, the Mercalli Scale provided significant historical information on earthquakes. It helps to better assess seismic hazards by providing information on how past earthquakes impacted areas.

Key Takeaways

The impact and destruction of an earthquake at specific locations are measured by the Mercalli Intensity Scale. By offering a more comprehensive view of an earthquake's effects on the environment and civilization, it enhances magnitude scales. For emergency management, engineering, and earthquake preparation, it is essential to understand both magnitude and intensity.

Seismic Wave Propagation and Earth's Interior

Wave Behavior:

A. **Reflection:** The phenomenon that occurs when seismic waves reflect off of a border between two distinct materials.
B. **Refraction:** When seismic waves bend as they travel through materials of varying densities, this phenomenon is known as Refraction.
C. **Absorption:** When seismic energy is absorbed, reducing the amplitude of the waves.
D. **Seismic Discontinuities:** The study of how seismic waves change speed and direction at different depths has led to the identification of major boundaries within the Earth's interior.

Internal Structure of the Earth

The Earth's interior is divided into three main layers: the crust, mantle, and core. Each layer has distinct physical and chemical properties.

The Crust: The crust of the Earth is the outermost layer of solid material that is made up of rocks and minerals. The crust of the Earth is the layer that is the thinnest of all the layers and varying in thickness that makes up the Earth.

Types of Crust:

A. **Continental Crust**: The average thickness of continental crust is30-70. It is primary composed of granitic rocks which are rich in silica and aluminum. Its density is lower than oceanic crust, averaging about 2.7 gm/cm^3. The Age of continental crust isolder than oceanic crust, with some regions being over 4 billion years old.
B. **Oceanic Crust**: The average thickness of Oceanic Crust is 5-10 kilometers. It is primary Composed of basaltic rocks, rich in iron and magnesium. Its density is higher than continental crust, averaging about 3.0 g/cm^3. It is younger than continental crust, with the oldest oceanic crust being about 200 million years old.

Mohorovicic discontinuity (Moho): Seismic waves exhibit significant variations in velocity near the boundary between the crust and the mantle, which is commonly referred to as the Mohorovicic discontinuity (Moho). It was determined that this boundary was established by a significant increase in seismic wave velocities (P-wave), which indicated a change in composition from crustal rocks to mantle rocks.

The Mantle: The mantle lies beneath the crust and extends to a depth of about 2,900 kilometers. It is composed of silicate minerals rich in iron and magnesium.

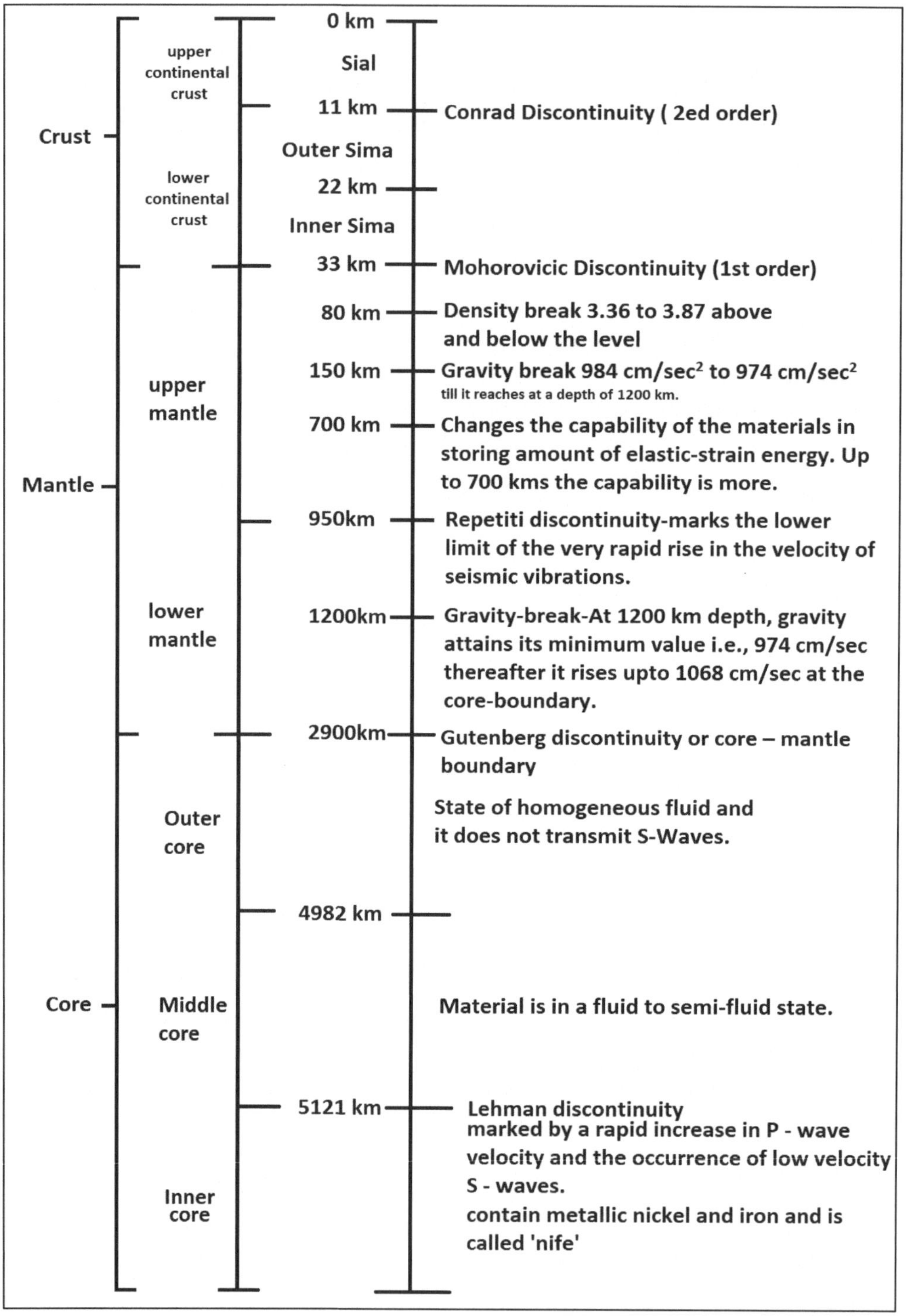

Figure 11.4- The Internal Structure of the Earth

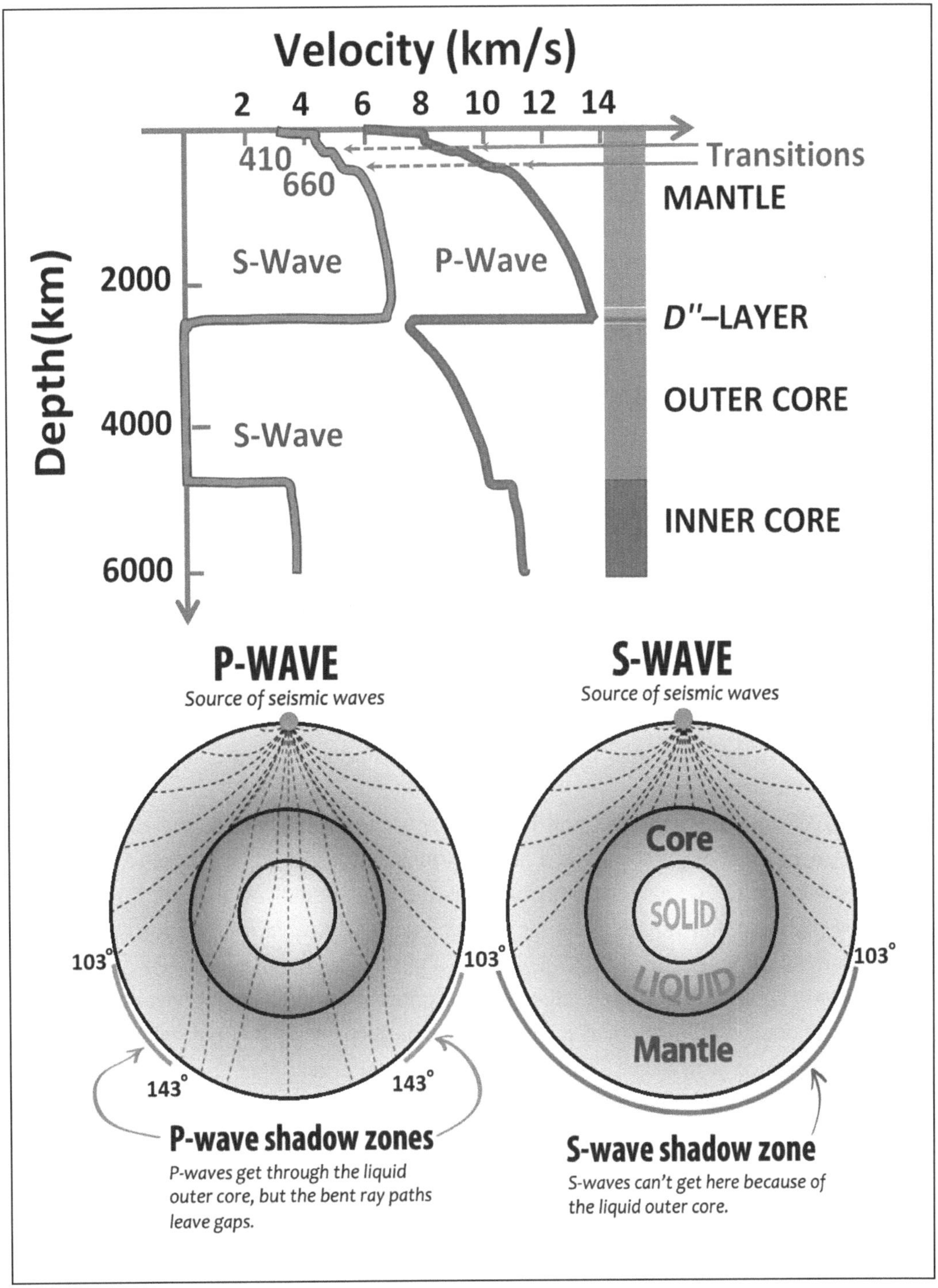

Figure 11.5: Seismic Wave Propagation and Earth's Interior

Seismic waves reveal the plastic asthenosphere, the transition zone, and variations in mineral structures. Alterations in wave velocities in the transition zone are indicated of changes in the phase and composition of the minerals.

Subdivisions of the Mantle:

A. **Upper Mantle**: It is extends from the base of the crust (33 km) to about 660 kilometers depth and includes the lithosphere (rigid outer part) and the asthenosphere (partially molten layer that allows tectonic plates to move). The asthenosphere is a ductile zone that is situated below the lithosphere and is responsible for facilitating the movement of plates tectonic.

B. **Transition Zone**: It is situated between 410 and 660 kilometers depth. High-pressure mineral phases, such as wadsleyite and ringwoodite, which are denser than minerals found in the upper mantle, are distinguish in this region from the upper mantle lithosphere.

C. **Lower Mantle**: It is extends from 660 kilometers to 2,900 kilometers. High-pressure minerals such as bridgmanite and ferropericlase, which are able to survive in the extreme conditions of the lower mantle, are the components that made up the lower mantle.

a. **D'' Layer**: The D" Layer is the border zone between the mantle and the outer core, and it is distinguished by the complex seismic wave behavior that that it displays.

b. **Composition**: Silicate minerals, particularly olivine, pyroxenes, and garnet, are the predominant minerals in the upper mantle, whereas perovskite and ferropericlase are the predominant minerals in the lower mantle. Peridotite is the most common rock type in the upper mantle.

c. **Convection**: Mantle convection is the slow, churning motion of the Earth's mantle caused by heat from the core. Plate tectonics is driven by the mantle convection process, which is also responsible for the movement of tectonic plates. This movement is what causes phenomena such as active volcanoes, earthquakes, and the formation of mountains.

d. **Gutenberg Discontinuity:** Gutenberg discontinuity is boundary between the mantle and the outer core. It is marked by a dramatic decrease in P-wave velocities and the disappearance of S-waves, indicating a liquid outer core.

The Core: There is a significant difference in wave velocities at the core-mantle boundary (CMB), which is located around 2,900 kilometers deep. The core is the deepest layer of the Earth, and it is mostly made of iron and nickel.

Lehmann Discontinuity:

The border between the outer core and the inner core is known as the Lehmann discontinuity. This phenomenon is characterized by an increase in the P-wave velocities that occur inside the solid inner core. The liquid outer core has a significant slowing effect on P-waves, and S-waves are unable to move through it, which results in the formation of shadow zones. P-waves rapidly accelerate once again inside the solid inner core, which is evidence that the inner core is solid.

The core is divided into two parts: the outer core and the inner core.

A. **Outer Core**: It is extends from 2,900 kilometers to about 5,150 kilometers. The state of outer core is liquid, composed mainly of iron and nickel, with lighter elements like sulfur and oxygen. S-waves are unable to travel through the liquid outer core, which results in the formation of a shadow zone that provides seismologists with information on the size and condition of the core. The geodynamo mechanism is responsible for the generation of the magnetic field of the Earth. This field is generated by the flow of molten iron and nickel in the outer core.

B. **Inner Core**: It is extends from about 5,150 kilometers to the Earth's center at 6,371 kilometers. The state of inner core is solid, due to the immense pressure that forces iron and nickel to crystallize despite the high temperatures. It is primarily Compose of iron, with some nickel and trace amounts of lighter elements. It is clear that the inner core is solid since the velocity of P-waves increases as they go deeper into the core. Inner core mainly composed of Iron (Fe)~85% by weight and Nickel (Ni): ~5-10% by weight. Approximately 12.8 to 13.0 grams per cubic centimeter is the average density of the inner core, which is the densest portion of the Earth.

Shadow Zones: Shadow zones are areas on the Earth's surface where no direct seismic waves from a particular earthquake are detected. These zones provide critical evidence about the Earth's internal structure.

P-wave Shadow Zone: It is between 103° and 142° from the earthquake source. Due to the refraction of P-waves at the boundary between the core and the mantle, this indicates a change in the composition of the material from the mantle to the liquid outer core.

S-wave Shadow Zone: It is beyond 103° from the earthquake source. S-waves do not pass through the liquid outer core, which results in the formation of a shadow zone which provides evidence of the liquid condition of the outer core.

A revolutionary change in our knowledge of the interior structure of the Earth has been brought about by seismology. By measuring seismic waves, scientists have been able to locate the crust, mantle, and core of the Earth, which has helped them to gain insight into the composition, condition, and dynamic processes that occur inside the Earth. This knowledge is essential for understanding geological processes, predicting earthquakes, and exploring natural resources.

Chapter 12
Earth's Thermal Structure, Internal Heat and Radioactivity of the Earth

The distribution of temperature in the Earth's interior is referred to as its thermal structure. Understanding the thermal structure of the Earth is crucial for understanding geodynamic processes including plate tectonics, mantle convection, and the formation of the Earth's magnetic field. Temperatures increase with depth, reflecting the substantial differences in the Earth's thermal structure from the surface to core.

1. Sources of Earth's Internal Heat

The Earth's internal heat is derived from several sources:

A. **Primordial Heat:** The term "primordial heat" refers to the heat that was left behind after the Earth was formed around 4.5 billion years ago. The accretion of planetesimals and the subsequent differentiation of the Earth's interior are the processes that may be responsible for its existence. It plays a crucial role in the early history of the Earth since it was responsible for supplying the first high temperatures that were required for geological processes.

B. **Radiogenic Heat:** Radiogenic heat is generated when unstable isotopes in the Earth undergo radioactive decay, which results in the heat being produced. During the process of decay, these isotopes emit particles and energy, which lead to the release of heat. It is a significant ongoing heat source that supports Earth's thermal processes across geological time.

C. **Tidal Heating:** The frictional forces that are associated with the gravitational interactions between the Earth, Moon, and Sun are responsible for the generation of tidal heating. Tidal heating, while very insignificant relative to other heat sources, contributes to certain areas, particularly in the Earth's oceans.

D. **Heat from Core Formation:** In the early phases of the creation of the Earth, the separation of the core from the mantle resulted in the release of a significant amount of heat. This was caused by the latent heat inherent within the crystallization process.

E. **Gravitational Heat:**Gravitational Heat generated by the compression and differentiation of materials within the Earth during its early formation. Gravitational potential energy is converted into heat as denser materials sink towards the center of the planet. In the early stages of Earth's history, it played a crucial role in the first heating and differentiation.

F. **Latent Heat:** When the Earth's interior undergoes phase transitions, latent heat is either released or absorbed. The inner core's crystallization and the mantle's material melting are two important examples. Latent heat contributes to Earth's thermal budget, especially in areas that are going through phase transitions.

2. **Temperature Distribution within the Earth:** The geothermal gradient is the term used to describe the increase in temperature that occurs when one goes deeper into the Earth. However, the rate of temperature increase is not uniform throughout the Earth's interior.

A. **The Crust**

a. **Temperature Range:** There are a number of factors that influence the temperature of the Earth's crust, including location and depth. Temperatures may vary from 0 degrees Celsius to over 200 degrees Celsius in the top few kilometers of the atmosphere.

b. **Geothermal Gradient:** Geothermal gradients in the crust may range from as low as 15 degrees Celsius per kilometer in stable, ancient cratonic regions to as high as 50 degrees Celsius per kilometer in tectonically active places. In continental regions, the geothermal gradient in the crust ranges between 25 and 30 degrees Celsius per kilometer. Because of its thinner nature and the existence of hydrothermal circulation, the geothermal gradient is often larger in the oceanic crust.

B. **The Mantle**

a. **Temperature Range:** At a depth of about 2,900 kilometers, the mantle stretches from the base of the crust to the depths of the earth. Temperatures in the upper mantle begin at roughly 500 degrees Celsius to 900 degrees Celsius close to the

Moho discontinuity and rise to approximately 1,600 degrees Celsius to 2,200 degrees Celsius at the base of the mantle, close to the border between the core and the mantle.

b. Temperature Distribution:

- **Upper Mantle**: The temperatures in the upper mantle continue to rise as one move deeper into the earth. With temperatures reaching from 1,300 to 1,500 degrees Celsius, the asthenosphere, which is a partly molten layer located inside the upper mantle, is able to display ductile behavior and facilitate the movement of tectonic plates.
- **Transition Zone**: In the transition zone, which is situated between 410 and 660 kilometers deep, there is a rise in both pressure and temperature, which causes changes in the mineral structures. For instance, olivine undergoes transformations into wadsleyite and ringwoodite. Temperatures might climb anywhere from 1,500 to 1,800 degrees Celsius.
- **Lower Mantle**: Temperatures in the lower mantle continue to rise with depth, reaching a maximum of around 2,200 degrees Celsius at the core-mantle border. Temperatures in the transition zone range from 1,800 degrees Celsius. The lower mantle is made up of minerals such as bridgmanite and ferropericlase, both of which are able to maintain their stability at these very high temperatures.

C. The Core

a. Outer Core: The outer core is a liquid layer that extends from about 2,900 kilometers to 5,150 kilometers in depth. It is mostly made of iron and nickel. At the border between the core and the mantle, temperatures in the outer core vary from roughly 3,200 to 4,000 degrees Celsius, rising to around 5,000 to 5,700 degrees Celsius at the boundary with the inner core. As a result of the high temperatures, the outer core is in a liquid condition. This is because the high temperatures prohibit iron and nickel from becoming solid.

b. Inner Core: There is a solid inner core that extends from about 5,150 kilometers to the center of the Earth, which is located at a depth of 6,371 kilometers. The inner core is mostly made of iron and some nickel. Temperature Range: It is thought that the temperature in the inner core ranges from around 5,500 degrees Celsius to 6,000 degrees Celsius. Despite the extreme temperatures, the iron is compelled to crystallize due to the great pressure that exists at this level (more than 3.5 million atmospheres), which results in the formation of a solid inner core.

1. Heat Transfer Mechanisms within the Earth

A number of different methods of heat transport are responsible for regulating the distribution of heat across the Earth:

A. Conduction: Direct contact between particles allows for the transmission of heat. Conduction is the primary mode of heat transport in the crust and lithosphere of the

Earth, which results in a gradual rise in temperature as one move deeper into the layers. The thermal conductivity of the rocks, which changes with composition and temperature, is known to be a significant factor in determining the effectiveness of conduction.

B. **Convection**: Heat transmission occurs when liquids or deformable solids move. Convection is the main method of heat transport in the mantle, where colder, denser material sinks and hotter, less dense material rises. Plate tectonics is driven by convection in the mantle, which also causes mountain formation, volcanic activity, and continent migration. Convection is another feature of the outer core that, in conjunction with Earth's rotation, produces the magnetic field via the geodynamo process.

C. **Radiation**: Heat transfer through electromagnetic waves. While radiation is significant in the transfer of heat from the Earth's surface to space, it is less important within the solid Earth, where conduction and convection dominate.

D. Radioactivity and Its Role in Earth's Internal Heat

a. Uranium-238 (U-238)

- Decay Series: Decays to lead-206 (Pb-206) via a sequence of stages.
- Half-life: 4.468 billion years, roughly.
- Heat Production: It is a significant source of radiogenic heat because of its lengthy half-life and widespread presence in the crust of the Earth.

b. Uranium-235 (U-235)

- Decay Series: Decays to lead-207 (Pb-207) via a sequence of stages.
- Half-life: 703.8 million years, roughly.
- Heat Production: As a result of its shorter half-life, it contributes to radiogenic heat, but to a smaller level than U-238 does.

c. Thorium-232 (Th-232)

- Decay Series: Decays to lead-208 (Pb-208) via a sequence of stages.
- Half-life: 14.05 billion years, roughly.
- Heat Production: It is a significant source of heat because of its abundance in the mantle and lengthy half-life.

d. Potassium-40 (K-40)

- Decay Modes: Breaks down into calcium-40 (Ca-40) and argon-40 (Ar-40).
- Half-life: 1.248 billion years, roughly.
- Heat Production: contributes to radiogenic heat, which is particularly important in the upper mantle and crust.

D. Mechanisms of Heat Generation

a. Alpha Decay

- Process: Release of an alpha particle from a radioactive atom's nucleus, consisting of two protons and two neutrons.
- Energy Release: Contributes to internal heat by releasing a substantial quantity of energy.

b. Beta Decay

- Process: Conversion of a proton into a neutron with the emission of a positron or a neutron into a proton with the emission of an electron (beta particle).
- Energy Release: Heat is produced, but it is often less than that produced by alpha decay.

c. Gamma Decay

- Process: Emission of photons with high energy, often known as gamma rays, from a nucleus that is radioactive.
- Energy Release: The heat that is produced by alpha and beta decay is increased by this process, but it is often of less significance.

E. Distribution of Radioactive Elements

a. Crust

- Concentration: In comparison to the mantle and the core, its radioactive element concentrations are much higher.
- Heat Production: Increased heat output per volume because to the high levels of Th, K, and U.

b. Mantle

- Concentration: Despite having lower levels of radioactive elements, it makes up a significant portion of the Earth's total radiogenic heat because of its size.
- Heat Production: Contributes significantly to maintaining plate tectonics and mantle circulation.

c. Core

- Concentration: Very low levels of radioactive elements.
- Heat Production: Latent heat and primordial sources provide for the majority of interior heat, with radiogenic sources contributing very little.

F. Thermal Anomalies and Hotspots

a. Hotspots:

- Localized volcanic activity occurs in hotspots, which are areas of the mantle where plumes of hot material ascend towards the surface.
- Iceland, Yellowstone, and the Hawaiian Islands are a few examples.
- Hotspots may provide information about deep mantle processes and are linked to greater than normal temperatures in the mantle.

b. Thermal Anomalies:

- Areas where temperatures diverge from the anticipated geothermal gradient are known as thermal anomalies.

- The existence of fluids, changes in the composition of the mantle, or previous volcanic and tectonic activity may all contribute to these abnormalities.
- Mid-Ocean Ridges: The upwelling of heated mantle material causes high temperatures in areas where new oceanic crust is created.

G. Geothermal Energy

a. Geothermal Energy: The heat produced by the Earth's core thermal structure is known as geothermal energy. Particularly in areas with strong geothermal gradients, like Iceland, New Zealand, and portions of the United States, it is used for heating, power production, and other purposes.

b. Geothermal Gradients: In areas with steep geothermal gradients, where temperature rises quickly with depth, geothermal energy potential is greater.

H. Implications of Earth's Thermal Structure

a. Mantle Convection and Plate Tectonics:

- Tectonic plate movement is driven by mantle convection, which is fueled by the Earth's thermal structure.
- During this process, new crust is formed at mid-ocean ridges, crust is recycled into the mantle at subduction zones, and mountains, earthquakes, and volcanoes are formed. Additionally, this process is responsible for the construction of submarines.

b. Magnetic Field Generation:

- The convecting molten iron in the outer core's thermal structure is essential for the creation of the Earth's magnetic field.
- The magnetic field's intensity and behavior may be affected by changes in the core's thermal structure.

c. Thermal Evolution of the Earth:

- Despite the fact that the Earth has been cooling ever since it was formed, radiogenic heat generation and tidal heating continue to contribute to the planet's internal energy budget.
- Geoscientists can forecast future geodynamic activity and simulate the planet's thermal development by having a thorough understanding of the Earth's thermal structure.

The Earth's thermal structure is extremely stratified, with temperatures increasing as one move deeper inside the globe. Its interior heat is derived from residues of its creation, radioactive decay, and tidal forces. Radioactive decay of isotopes like as uranium, thorium, and potassium is especially important in creating heat, which drives geological processes like mantle convection and plate tectonics. This mix of heat sources keeps the Earth geologically active, resulting in phenomena like earthquakes, volcanoes, and the Earth's magnetic field.

Chapter 13
Earth's Gravity and Magnetic Field

Introduction

The Earth's gravity and magnetic field are fundamental forces shaping our planet's physical environment and influencing various geological and atmospheric processes. The interior dynamics, structure, and interactions of Earth with space can all be better understood by knowing these forces.

Earth's Gravity

1. Fundamentals of Gravity

A. Definition: Gravity is the force that attracts two objects together. It gives physical items weight on Earth, forcing them to fall toward the ground.

B. Newton's Law of Universal Gravitation: Claims that there is an attraction between all masses. The square of the distance (r) between the centers of two masses (m1 and m2) is inversely proportional to the force (F) between them, which is proportional to the product of their masses:

$$F = G\frac{m^1 * m^2}{r^2}$$

where G is the gravitational constant.

2. Earth's Gravitational Field

A. Gravitational Acceleration: On Earth's surface, the acceleration due to gravity (g) is roughly 9.8 m/s^2, though it varies slightly because of the planet's rotation, shape, and internal density changes.

B. Geoid: The Earth's theoretical shape, which represents the average sea level worldwide, is impacted by gravitational imbalances brought on by variations in crust and mantle density.

C. Equipotential Surface: Surfaces that have a constant gravitational potential. The example of an equipotential surface is the geoid.

3. Gravity Anomalies

A. Bouguer Anomaly: provides information about subsurface structures by adjusting recorded gravity to take topography and elevation into awareness.

B. Free-Air Anomaly: Considering elevation but not topography, it calculates the difference between the observed and theoretical gravity at sea level.

4. Applications of Gravity Studies

A. Geological Exploration: Gravity surveys are useful for locating subsurface features such tectonic features, mineral deposits, and oil reservoirs.

B. Geodesy: The shape of the Earth and changes in sea level are determined in part by precise gravity measurements.

C. Climate Studies: Changes in hydrological cycles and the loss of ice mass can be observed by satellite measurements of gravity variations.

Earth's Magnetic Field

1. Fundamentals of Magnetism

A. Magnetic Field: A vector field that surrounds electric currents and magnetic materials and applies force to charged particles and other magnetic objects.

B. Earth's Magnetic Field: Formed mostly by the geodynamo process, this moves molten iron in the outer core.

2. Structure of Earth's Magnetic Field

A. Dipole Field: The Earth has a magnetic north and south pole, which makes it resemble a bar magnet. The tilt of the magnetic axis relative to the rotating axis is approximately 11 degrees.

B. Magnetosphere: The region around Earth where its magnetic field interacts with the solar wind, protecting the planet from charged particles.

3. Geodynamo Theory

A. Generation: Electric currents are produced by convection currents in the liquid outer core, which are driven by heat from radioactive decay and leftover heat from the formation of Earth. Magnetic fields are generated by these currents.

B. Sustaining Mechanism: The interaction between the fluid motion and magnetic fields maintains the geodynamo.

4. Magnetic Anomalies

A. Regional Anomalies: Variations in the magnetic field caused by local differences in Earth's crust, such as mineral composition and geological structures.

B. Magnetic Reversals: A chronology of geomagnetic history is provided by the orientation of magnetic minerals in rocks, which records periodic reversals of Earth's magnetic poles.

5. Applications of Magnetic Studies

A. Navigation: Using the Earth's magnetic field as a guide, the magnetic compass has been an essential navigational aid for generations.

B. Mineral Exploration: Magnetic surveys detect ore bodies and geological structures based on their magnetic properties.

C. Plate Tectonics: Studies of magnetic anomalies on the ocean floor provide evidence for seafloor spreading and plate tectonic theory.

D. Space Weather: Understanding the magnetosphere makes it easier to forecast space weather events that could have an impact on power grids, satellites, and communication networks.

Interaction between Gravity and Magnetic Field

1. Combined Geophysical Surveys

A. Integrated Approach: Subsurface geology can be better understood by combining magnetic and gravity investigations.

B. Data Interpretation: Identification of structural features and differentiation of various rock types are facilitated by the correlation of magnetic and gravity data.

2. Earth's Interior Structure

A. Core-Mantle Boundary: Changes in magnetic and gravitational fields reveal information about the dynamics and structure of the core-mantle interface.

B. Mantle Convection: Mantle convection patterns and their function in plate tectonics are better understood because of gravity anomalies and magnetic data.

3. Monitoring Environmental Changes

A. Sea Level Change: Satellites measure gravity, which measures variations in ice mass and sea level.

B. Geodynamic Processes: Geodynamic processes like earthquakes and volcanic activity are tracked by changes in the magnetic and gravitational fields.

Advanced Technologies in Gravity and Magnetic Studies

1. Satellite Missions

A. GRACE (Gravity Recovery and Climate Experiment): Monitors shifts in the distribution of mass, including ice sheets and groundwater, by measuring variations in Earth's gravitational field.
B. GOCE (Gravity Field and Steady-State Ocean Circulation Explorer): Provided high-resolution gravity data to refine the geoid.

2. Ground-Based Surveys

A. Gravimeters: Geological and environmental investigations use instruments that monitor minute changes in gravity.
B. Magnetometers: Magnetometers are very important for magnetic studies and navigation because they measure the strength and direction of magnetic fields.

3. Computational Modeling

A. Geophysical Models: Numerical models that use geological data to estimate how magnetic and gravitational fields behave.
B. Data Integration: Generating detailed models of Earth's interior and surface processes by integrating data from various sources (e.g., satellites, ground surveys).

It is essential for a variety of scientific disciplines, including environmental science, space studies, and geophysics, to understand the gravity and magnetic field of Earth. The examination of these fields not only offers practical applications in resource exploration, environmental monitoring, and hazard mitigation, but also reveals insights into the internal structure and dynamics of the planet. Our comprehension of these fundamental forces that influence our world will be further enhanced by ongoing developments in computational methods and technology.

Chapter 14
Earthquakes

Introduction to Earthquake

Earthquakes are one of the most powerful natural events that can happen. They cause the ground to shake quickly because energy is released from the Earth's base. Their results can be very bad for people, buildings, and the environment. Understanding the causes of earthquakes, how they happen, and where they happen is important for lessening their affects and getting ready for them.

Causes of Earthquakes

A. Tectonic Earthquakes

Plate Tectonics Theory

a. Description: The tectonic plates that make up the Earth's lithosphere float on the semi-fluid asthenosphere underneath them.

b. Movement: When these plates touch at their borders, stress builds up and is finally released as earthquakes.

Types of Plate Boundaries

a. Divergent Boundaries: When plates separate, strain is created that can lead to earthquakes. The Mid-Atlantic Ridge is one example.

b. Convergent Boundaries: Compressional forces are created when plates move in the direction of one another and one is pressed beneath the other. The Pacific Ring of Fire is one example.
c. Transform Boundaries: Shear tension is created when plates glide past one another horizontally. The San Andreas Fault in California is one example.

B. Volcanic Earthquakes

a. Description: Earthquakes may be caused by magma movement in the Earth's crust.
b. Mechanism: Pressure builds up in the surrounding rock as magma rises, resulting in seismic activity and fracturing.
c. Examples: Earthquakes around active volcanoes like Mount St. Helens (USA).

C. Human-Induced Earthquakes

a. Description: Earthquakes can be caused by human activities including mining, hydraulic fracturing (fracking), geothermal energy extraction, and reservoir-induced seismicity.
b. Mechanism: Seismic occurrences are caused by these actions, which change the tension and pressure conditions in the Earth's crust.
c. Examples: Increased seismic activity in Oklahoma due to fracking, earthquakes near large reservoirs like the Koyna Dam in India.

D. Other Causes

a. Collapse Earthquakes

- Description: Earthquakes brought on by mines or caves collapsing underground.
- Examples: Collapses in karst regions and mining-induced collapses.

b. Explosion Earthquakes

- Description: Explosions, such nuclear tests or massive chemical explosions can produce seismic waves.
- Examples: Nuclear tests in Nevada (USA) and North Korea.

Effects of Earthquakes

1. Primary Effects

A. Ground Shaking

a. Description: Vibrations caused by the release of seismic energy.
b. Impact: Can cause buildings, bridges, and other structures to collapse.

B. Surface Rupture

a. Description: A fault line is where the Earth's surface breaks and shifts.
b. Impact: Damages infrastructure creates fissures and faults visible on the surface.

2. Secondary Effects

A. Landslides

a. Description: There is a possibility that earthquakes can cause landslides in areas that are steep or mountainous.

b. Impact: Buries communities, blocks roads and rivers, causing additional damage.

B. Tsunamis

a. Description: Underwater earthquakes can displace large volumes of water, generating tsunamis.

b. Impact: Devastates coastal areas, causing loss of life and property.

C. Liquefaction

a. Description: Intense shaking causes saturated soil to momentarily lose its strength and behave like a liquid.

b. Impact: Causes buildings to sink or tilt, damaging foundations.

D. Fires

a. Description: Gas lines that have ruptured, electrical short circuits, and water pipes that have broken are all potential causes of fires.

b. Impact: Causes extensive damage, particularly in urban areas.

3. Social and Economic Effects

A. Casualties and Injuries

a. Description: Earthquakes have the potential to cause a substantial number of fatalities and injuries.

b. Impact: Overwhelms healthcare systems, lead to long term trauma.

B. Economic Losses

a. Description: In addition to the destruction of property and infrastructure, there is also an interruption of economic activity.

b. Impact: Can lead to billions of dollars in damages and long-term economic downturns.

C. Displacement

a. Description: As a result of the potential for aftershocks or damage, individuals can be compelled to evacuate their residences.

b. Impact: Leads to temporary or permanent displacement and the creation of refugee camps.

D. Psychological Effects

a. Description: Earthquakes have the potential to induce a substantial amount of psychological stress and trauma.

b. Impact: Long-term mental health issues and PTSD.

Measuring Earthquakes

It is essential to measure earthquakes in order to understand their properties, forecast future occurrences, and lessen their effects.

Seismometers and Seismographs

1. Seismometers

A. Description

a. Definition: When it comes to detecting and measuring the motion of the ground, a seismometer is an instrument that is capable of detecting and measuring the vibrations that are created by seismic waves that are generated by earthquake

b. Components: In general, it is composed of a mass, which is also referred to as a proof mass or inertial mass, which is hung by a spring and a damping system in order to maintain stability and prevent oscillation.

B. Working Principle

a. Inertia: Because of the force of inertia, the mass does not move when the ground moves, but the frame moves along with the ground.

b. Detection: In order to transform the relative motion between the mass and the frame into electrical signals, the motion is recorded.

C. Types of Seismometers

a. Vertical Seismometers: Determine the vertical motion of the ground.

b. Horizontal Seismometers: Measure horizontal ground motion in one or two perpendicular direction.

2. Seismographs

A. Description

a. Definition: The seismograph is a comprehensive system that combines a seismometer, a recording device, and a timekeeping mechanism. Its purpose is to record the motion of the ground over a period of time.

b. Purpose: A seismogram, which is an illustration of seismic waves, is produced in order to provide results.

B. Seismograms

a. Components: The display of the arrival times and amplitudes of surface waves, P waves, and S waves is of great importance.

b. Usage: In order to investigate the features of seismic waves, it is necessary to identify the epicenter and determine the magnitude of earthquakes.

C. Modern Seismic Networks

a. Global Seismographic Network (GSN)

- Description: A network consisting of seismographic stations of superior quality that are spread out internationally.
- Purpose: In order to keep track of seismic activity on a worldwide scale and supply information for the purpose of earthquake research and risk assessment.

b. Regional Seismic Networks

- Description:Networks of seismographic stations located within certain regions, with the purpose of providing extensive monitoring of seismic activity in the local area.
- Examples: Southern California Seismic Network (SCSN), Japan Meteorological Agency (JMA) network.

3. Magnitude Scales

A. **Richter Scale**

a. **Development**

- Inventor: Charles F. Richter in 1935.
- Purpose: The purpose of this is to determine the magnitude of earthquakes by analyzing the amplitude of seismic waves.

b. **Calculation**

- Amplitude: Seismographs are used to measure the maximum amplitude of seismic waves that are recorded on the instrument.
- Logarithmic Scale: With each whole number increase, the amplitude increases by a factor of 10, and the amount of energy released increases by approximately 31.6 times.

c. **Limitations**

- Local Magnitude: This method is effective for earthquakes of small to medium magnitude, but it is less accurate for earthquakes that are huge and remote.

B. **Moment Magnitude Scale (Mw)**

a. **Development**

- Inventors: Thomas C. Hanks and Hiroo Kanamori in the late 1970s.
- Purpose: In order to provide a more precise measurement of the magnitude of an earthquake, particularly for significant occurrences.

b. **Calculation**

- **Seismic Moment (M_0):** The region of the fault that slipped, the average slip, and the stiffness of the rocks are all factors that are taken into consideration.
- **Formula:**

$$M_w = \frac{2}{3}\log_{10}(M_0) - 10.7$$

c. **Advantages**

- **Accuracy:** As a result of being more precise for large earthquakes, it offers a more exact measurement of the total energy released.
- **Consistency:** Can be used for all sizes of earthquakes globally.

C. **Other Magnitude Scales**

a. **Body-Wave Magnitude (Mb):** The amplitude of body waves, including P and S waves, can be measured, which is beneficial for earthquakes that are deeper.

b. **Surface-Wave Magnitude (Ms):** This device is used to determine the amplitude of surface waves. Useful for earthquakes that are quite large and shallow.

c. **Duration Magnitude (Md):** According to the length of time that the shaking was captured on seismograms. It is frequently employed for earthquakes that are smaller and more localized.

4. **Intensity Scales**

A. **Modified Mercalli Intensity Scale (MMI)**

a. **Description**

- Definition: An evaluation scale that is qualitative in nature that measures the effects and damage that an earthquake causes.
- Scale: Ranges from I (not felt) to XII (total destruction).

b. **Assessment**

- Criteria: Based on observations of people, buildings, and natural features.
- Usage: Useful for assessing the impact of an earthquake on communities.

c. **Advantages**

- Human Impact: It is a measurement that provides an assessment of how individuals see an earthquake and the level of the damage that it causes.
- Historical Earthquakes: It has the capability to be utilized for historical events for which there is a lack of available instrumental data.

B. **European Macroseismic Scale (EMS-98)**

a. **Description**

- Definition: There is a qualitative scale that is comparable to the MMI, but it has been specifically built for usage in Europe.
- Scale: Ranges from I (not felt) to XII (completely devastating).

b. **Assessment**

- Criteria: This is based on comprehensive rules for evaluating the damage to buildings, the consequences on the ground, and human perception.

5. **Locating Earthquakes**

A. **Epicenter and Hypocenter**

a. **Epicenter**

- Definition: A place on the surface of the Earth that is directly above the focal point of the earthquake, also known as the hypocenter.
- Significance: This refers to the location where the impacts of an earthquake are typically the most destructive.

b. **Hypocenter**

- Definition: This refers to the precise location within the Earth's surface where the earthquake manifests itself.
- Depth: Can range from shallow (less than 70 km) to deep (more than 300 km).

B. **Triangulation Method**

a. **Principle**

- Description: Each seismograph station is used to determine the distance from the epicenter by utilizing the arrival times of P and S waves at the various seismograph stations.

- Process: We may determine the epicenter of the earthquake by drawing circles with radii that are equal to the distances from each station to the epicenter. The location where these circles connect is the epicenter of the earthquake.

b. Steps

- Determine Arrival Times: At each station, it is necessary to measure the difference in arrival timings between the P waves and the S waves.
- Calculate Distances: Use the travel-time difference to calculate the distance to the epicenter.
- Triangulate: Plot the distances on a map to find the intersection point.

C. Advanced Techniques

a. Seismic Tomography

- Description: It is comparable to medical CT scans in that it creates three dimensional images of the interior of the Earth by using seismic waves.
- Usage: Allows for a more precise localization of earthquakes and facilitates the investigation of the structure of the Earth's interior.

b. InSAR (Interferometric Synthetic Aperture Radar)

- Description: InSAR is use in satellite radar technology to identify the ground deformation that is a result due to earthquakes.
- Usage: The provision of extensive maps of surface displacement comes in handy while attempting to understand the movement of faults and the mechanics of earthquakes.

Earthquake Early Warning Systems

1. Principles

A. Detection

a. Initial Waves: It is important to detect the initial P waves, which travel at a faster speed and cause less damage. This allows for an early warning to be provided before the arrival of the S waves and surface waves, which are more damaging.

b. Sensors: In real time, networks of seismometers are able to detect and analyze data pertaining to seismic activity.

B. Warning

a. Communication: When a system detects something, it immediately sends notifications to both human users and automated systems.

b. Actions: Warnings have the ability to initiate safety procedures, which include the suspension of trains, the stopping of industrial activities, and the distribution of public warnings.

2. Examples of Early Warning Systems

A. Shake Alert (USA)

a. Description: An early warning system for earthquakes that is designed to be used on the West Coast of the United States.

b. Components: The program employs a network of seismometers, algorithms, and communication systems to identify earthquakes and provide alerts to the appropriate authorities

B. Earthquake Early Warning (EEW) System (Japan)

a. Description: It is a system that operates on a national scale and offers early warnings of earthquakes.

b. Features: Rapid detection and widespread dissemination of warnings via TV, radio, smart phones, and other channels.

3. Benefits and Challenges

A. Benefits

a. Life-Saving: Offers individuals the opportunity to take preventative measures within the vital seconds to minutes.

b. Infrastructure Protection: Facilitates the implementation of preventative steps by automated systems in order to safeguard infrastructure

B. Challenges

a. False Alarms: Maintaining a quick and accurate warning system while minimizing the occurrence of false alarms.

b. Coverage: When it comes to places with limited seismic networks, ensuring broad coverage is extremely important.

Earthquake Prediction and Preparedness

1. Prediction

A. Current Limitations

a. Description: The task of accurately forecasting the precise moment and location of earthquakes is a highly tough attempt.

b. Research: Studies that are currently being conducted on seismic patterns, animal behavior, and antecedents such as shock waves.

B. Probabilistic Forecasting

a. Description: This function provides an estimation of the probability of an earthquake occurring in a particular region within a specified time period.

b. Use: Helps in risk assessment and preparedness planning.

2. Preparedness

A. Building Codes

a. Description: Regulations that guarantee the ability of structures to withstand the effects of seismic forces.

b. Examples: Earthquake-resistant designs, retrofitting older buildings.

B. Emergency Plans

a. Description: During an earthquake, it is important to have strategies in place for evacuating, communicating, and allocating resources.

b. Components: Public education, drills, and training for first responders.

C. Early Warning Systems

a. Description: Technologies that are able to detect the first waves of seismic activity and deliver a brief warning before the main shaking begins.

b. Examples: Shake Alert in the USA, Earthquake Early Warning (EEW) in Japan.

Distribution of Earthquakes

1. Global Distribution

A. Earthquake Zones

a. Description: Tectonic plate borders are the locations where earthquakes are most likely to occur.

b. Major Zones: Pacific Ring of Fire, Alpine-Himalayan Belt, Mid-Atlantic Ridge.

B. Pacific Ring of Fire

a. Description: The region in the basin of the Pacific Ocean that is characterized by a significant number of earthquakes and volcanic eruptions to take place.

b. Characteristics: The presence of subduction zones, transform faults, and volcanic arcs is responsible for the high seismic activity.

c. Examples: Japan, Indonesia, Chile, and Alaska.

C. Alpine-Himalayan Belt

a. Description: From the region of the Mediterranean, it stretches all the way to Southeast Asia, passing via the Himalayas.

b. Characteristics:The collision of the African, Indian, and Eurasian plates has resulted in a very high level of seismic activity.

c. Examples: Earthquakes in Turkey, Iran, and Nepal.

D. Mid-Atlantic Ridge

a. Description: A diverging barrier that is located in the middle of the Atlantic Ocean and is responsible for the formation of new oceanic crust.

b. Characteristics: In the vicinity of Iceland and the Azores, there is a moderate level of seismic activity that can be attributed to the spreading of the seafloor.

2. Regional Distribution

A. North America

a. Western USA: The San Andreas Fault and other faults in California are seeing a high level of seismic activity recently.

b. Alaska: These earthquakes are caused by the subduction of the Pacific Plate beneath the North American Plate, which causes frequent earthquakes.

c. Eastern USA: Less frequent but possible intraplate earthquakes.

B. South America

a. Andean Region: Due to subduction, there is a significant amount of seismic activity along the Peru-Chile Trench.

b. Notable Earthquakes: 1960 Valdivia earthquake (magnitude 9.5), the largest recorded earthquake.

C. Asia

a. Japan: Subduction zones surrounding the Pacific Plate are responsible for the high seismic activity that occurs.

b. Indonesia: Frequent earthquakes due to complex tectonic interactions.

c. Himalayan Region: Earthquakes due to the collision between the Indian and Eurasian plates.

D. Europe

a. Mediterranean Region: Seismic activity due to the collision of the African and Eurasian plates.

b. Notable Earthquakes: 1908 Messina earthquake in Italy, 1755 Lisbon earthquake in Portugal.

E. Africa

a. East African Rift: The African Plate is moving away from one another, which is causing seismic activity.

b. North Africa: Earthquakes related to the collision between the African and Eurasian plates.

F. Oceania

a. New Zealand: High seismic activity along the Pacific and Australian Plate boundary.

b. Papua New Guinea**:** Frequent earthquakes due to complex tectonic interactions.

Strong natural occurrences, earthquakes are brought on by the release of energy from the Earth's crust. Their primary causes are tectonic, though they can also be caused by volcanic eruptions or human activity. Earthquakes can have catastrophic consequences, including long-term societal consequences, economic harm, and mortality. Assessing risk and putting preparedness measures in place are made easier with an understanding of seismic distribution. Even while it's still difficult to forecast earthquakes, technological and seismological advancements are making it easier for us to lessen their effects. We can lessen the likelihood of disaster by better protecting infrastructure and communities through earthquake research.

Chapter 15
The Formation of the Hydrosphere

1. Introduction

All of the water on Earth, including rivers, lakes, glaciers, groundwater, and atmospheric water vapor, is referred to as the hydrosphere. Understanding its genesis and history is essential to understanding the geology, climate, and evolution of life on Earth.

2. Origin of Earth's Water

A. Primordial Sources

a. Volcanic Outgassing: The great volcanic activity that occurred in the early Earth's history released water vapor and other gasses that had been trapped in the interior of the planet. The first pooling of water in the atmosphere was strongly influenced by this process, which contributed significantly.

b. Cometary and Asteroidal Delivery: Significant amounts of water were delivered to the early Earth by water-rich meteors that bombarded it. Through collisions, carbonaceous chondrite asteroids added to Earth's water inventory and also contained water.

B. Accretion and Differentiation

a. Accretion: The mass and heat of Earth were derived from the accretion of planets and other celestial bodies during its formation. During this process, water was integrated into the materials of the Earth.

b. Differentiation: Water was extracted from minerals and gathered on the surface as the Earth's core, mantle, and crust separated.

3. Cooling and Condensation

A. Atmospheric Cooling

a. Early Atmosphere Composition: Methane, carbon dioxide, water vapor, and other gases produced by volcanic outgassing predominate.

b. Cooling: As the Earth cooled, the water vapor in the atmosphere began to condense.

B. Formation of Liquid Water

a. Condensation Process: As the temperature of the atmosphere immersed below the boiling point of water, water vapor began to condense into liquid water.

b. Accumulation Process: The first bodies of water were formed when liquid water gathered on the surface of the earth.

4. Formation of Oceans and Other Water Bodies

A. Early Oceans

a. **Initial Formation Process**: Primordial oceans were created when condensed water accumulated in low-lying regions of the Earth's surface. Geological records show that a large portion of the Earth's surface was covered by oceans in the past.

b. **Chemical Evolution**:

- Interaction with Crust: The interaction of water with the crust of the Earth resulted in the dissolving and precipitation characteristics of minerals.
- Salinity: The salinity of ocean water arose with time as a result of the presence of dissolved minerals.

B. Rivers, Lakes, and Groundwater

a. **Surface Water**:

- Rivers and Lakes: The formation of lakes occurred as a result of the accumulation and movement of precipitation throughout the surface of the Earth, which resulted in the formation of river valleys and the filling of depressions.
- Watersheds: Regions that are clearly defined and contain surface water that drains into a common outlet, such as a river, lake, or ocean.

b. **Groundwater**:

- Infiltration: The precipitation penetrated the soil and the layers of permeable rock, which resulted in the formation of groundwater reservoirs and the replenishment of aquifers.
- Aquifers: It is the underground layers of water-bearing rock that are responsible for holding a substantial volume of groundwater.

C. Glaciers and Ice Caps (Cryosphere)

- **Formation**: The accumulation of precipitation in the form of ice and snow, which led to the formation of glaciers and ice caps, occurred in the Polar Regions and at higher altitudes, when temperatures remained below freezing.
- **Glacial Cycles**: The distribution and size of glaciers and ice caps were influenced by the periods of glaciation and interglacial eras that occurred respectively.

5. Evolution of the Hydrosphere

A. Geological and Climatic Influences

a. Plate Tectonics:

- Impact on Hydrosphere: The movement of tectonic plates had an effect on the distribution of ocean basins and continents, which contributed to changes in climatic patterns and ocean current direction.
- Hydrothermal Activity: The chemical evolution of the seas was influenced by the interaction between saltwater and the crust of the Earth at mid-ocean ridges.

b. Climate Change:

- Ice Ages: Glaciers that occurred on a regular basis had an effect on the levels of the oceans, the circulation of the oceans, and the distribution of water between the oceans, the atmosphere, and the ice sheets.
- Greenhouse Effect: Variations in the composition of the atmosphere, such as shifts in the levels of greenhouse gases, had an impact on the hydrological cycle and the temperatures of the entire planet.

B. Biological Influences

a. Emergence of Life:

- Role of Water: It was water that served as a medium for the biochemical events that were essential for the beginning and development of life.
- Photosynthesis: The chemistry of the hydrosphere and atmosphere was altered as a result of the production of oxygen by the earliest photosynthetic organisms.

b. Biogeochemical Cycles:

- Carbon Cycle: The climate and ocean chemistry were impacted by the carbon exchange between the hydrosphere, atmosphere, biosphere, and lithosphere.
- Nitrogen and Phosphorus Cycles: The hydrosphere is responsible for the circulation of essential nutrients, which in turn sustains both marine and terrestrial ecosystems.

6. Modern Hydrosphere

A. Components of the Hydrosphere

a. Oceans:

- Extent: They are responsible for around 71% of the Earth's surface and hold 97% of the water on the globe.

- Circulation: Wind, temperature differential, and salinity all play a role in the phenomena that influence the climate and weather patterns of the entire planet.

b. **Freshwater Systems**:

- Rivers and Lakes: Make available essential habitats for wildlife as well as water resources that can be utilized by humans.
- Groundwater: Provides water for drinking and irrigation, which is stored in aquifers that are located beneath the surface of the earth.

c. **Cryosphere**:Ice and Snow: Play an essential part in manipulating the temperature of the Earth and the level of the sea.

B. **Human Impact on the Hydrosphere**

a. **Water Use**:

- Consumption: Significant amounts of freshwater are extracted from the environment by means of agriculture, industry, and domestic use.
- Pollution: Urbanization, industrial waste, and agricultural runoff are all factors that contribute to water pollution, which in turn has an impact on ecosystems and the health of humans.

b. **Climate Change**:

- Global Warming: As temperatures continue to rise, the hydrological cycle is impacted, which results in alterations to the patterns of precipitation, the melting of glaciers, and an increase in sea levels.
- Ocean Acidification: The increased absorption of carbon dioxide by seas results in acidification, which has consequences for marine life and ecosystems.

There were a number of complex interactions that took place between geological, atmospheric, and biological processes during the development of the hydrosphere. Outgassing from volcanoes and impacts from comets, as well as the condensation of water vapor and the development of oceans, rivers, lakes, and glaciers, are all examples of the processes that have contributed to the formation of the water systems on Earth. One of the most important things to do in order to address present difficulties with water supplies, climate change, and environmental sustainability is to get an understanding of the genesis and evolution of the hydrosphere.

Chapter 16
The Formation of the Atmosphere

Introduction

Throughout the course of billions of years, the atmosphere of the Earth has developed into a system that is both dynamic and complicated. It is essential to have a solid understanding of its origins and growth in order to have a complete comprehension of the climate, weather, and conditions that are conducive to life on this planet.

1. Initial Atmosphere

Primary Atmosphere

A. **Composition**: **Hydrogen (H_2) and Helium (He)**: These light gases, which were available in large quantities in the early stages of the solar system, made up the majority of the primary atmosphere.

B. **Loss of Primary Atmosphere**:

Solar Wind: The lightweight hydrogen and helium atoms were removed by the powerful solar wind that was present in the newborn Sun.

Weak Gravity: The gravity of the Earth was not strong enough to hold these light gases from being carried away by the solar wind.

2. **Formation of the Secondary Atmosphere**

Volcanic Outgassing

A. **Process**:

a. Volcanic Activity: Early Earth was subjected to extensive volcanic activity, which resulted in the release of gasses that had been confined within the planet's interior.

b. Gases Released: Water vapor (H_2O), carbon dioxide (CO_2), methane (CH_4), ammonia (NH_3), nitrogen (N_2), and sulfur gases.

B. **Formation**: As a result of the accumulation of these gases, the secondary atmosphere was formed. This secondary atmosphere was significantly more stable and thicker than the primary atmosphere.

3. **Evolution of the Secondary Atmosphere**

A. **Cooling and Condensation**

Water Vapor Condensation:

a. Cooling Earth: Condensation of water vapor in the atmosphere resulted in the formation of liquid water when the planet cooled.

b. Formation of Oceans: A considerable amount of water vapor was removed from the atmosphere as a result of the accumulation of condensed water that led to the formation of the first seas.

B. **Chemical Reactions**

a. **Ammonia Breakdown:**

- Photodissociation: Ultraviolet (UV) rays emitted by the Sun facilitated the breakdown of ammonia (NH_3) into nitrogen (N_2) and hydrogen (H_2).
- Release of Nitrogen: The nitrogen content of the atmosphere was also raised as a result of this process.

b. **Methane and Carbon Dioxide**:

- Oxidation of Methane: The process of oxidizing methane (CH_3) resulted in the production of carbon dioxide (CO_2) and water (H_2O).
- Carbonate Formation: CO_2 was dissolved in water and then interacted with minerals to form carbonate rocks, which resulted in a decrease in the amount of CO_2 in the atmosphere.

4. **The Great Oxygenation Event**

A. **Emergence of Photosynthesis (Cyanobacteria)**

a. Photosynthesis: The ability to undertake oxygenic photosynthesis, which involves the utilization of sunlight to convert carbon dioxide (CO_2) and water into organic matter and oxygen (O_2), was developed by cyanobacteria.

b. Oxygen Production: As a result of this process, oxygen was released as a byproduct, which led to a gradual increase in the concentration of oxygen in the atmosphere.

B. **Accumulation of Oxygen**

a. **Initial Absorption**:

- Iron Oxidation: In the beginning, oxygen began to interact with iron in the oceans, resulting in the production of banded iron formations (BIFs). This interaction prevented iron from originally collecting in the atmosphere.

b. **Atmospheric Oxygen**:

- Saturation: On account of the oxidation of iron and other reactive substances, oxygen started to build in the atmosphere.
- Great Oxygenation Event (GOE): This occurrence took place approximately 2.4 billion years ago, indicating a significant increase in the levels of oxygen in the atmosphere.

5. Formation of the Ozone Layer

A. Production of Ozone

Oxygen and UV Radiation:

a. Photodissociation: Ultraviolet radiation has the ability to break down oxygen molecules (O_2) into little oxygen atoms (O).

b. Ozone Formation: The formation of ozone (O_2) occurred as a result of the reaction between these oxygen atoms and O_2.

B. Importance of the Ozone Layer

UV Protection:

a. Shielding Effect: The ozone layer is responsible for acting as a shield, shielding living creatures from the harmful effects of ultraviolet radiation that is emitted by the sun.

b. Enabling Terrestrial Life: As a result of this protection, life was able to flourish on land, as it decreased the likelihood of DNA damage caused by ultraviolet radiation.

6. Further Evolution of the Atmosphere

A. Biological Influence

Biogeochemical Cycles:

a. Carbon Cycle: A balance in the atmosphere is maintained through the regulation of CO_2 and O_2 levels through the processes of photosynthesis and respiration.

b. Nitrogen Cycle: Nitrogen-fixing bacteria have the ability to transform atmospheric nitrogen into forms that are biologically useful, so imposing an influence on the composition of the atmosphere.

B. Geological and Climatic Influences

Volcanic Activity:

a. Gas Emissions: As volcanic activity persisted, gases like carbon monoxide, sulfur dioxide, and water vapor were produced, which had an effect on the composition of the atmosphere.

b. Climate Impact: Large eruptions have the potential to release aerosols into the atmosphere, which can contribute to the global climate by reflecting sunlight and causing the globe to become cooler.

C. **Weathering and Carbon Sequestration**:

a. Chemical Weathering: As a result of the reaction between carbon dioxide and the minerals present in rocks, carbonates are formed and preserved in the crust of the Earth.

b. Long-term Climate Regulation: This mechanism serves as a long-term sink for carbon dioxide in the atmosphere, hence maintaining stable temperatures across the globe.

7. Modern Atmosphere

A. Composition

a. Major Components:

- Nitrogen (N_2): Approximately 78%
- Oxygen (O_2): Approximately 21%
- Argon (Ar): Approximately 0.93%
- Carbon Dioxide (CO_2): Approximately 0.04%
- Other Gases: Trace amounts of neon, helium, methane, krypton, and hydrogen.

b. Water Vapor:

Variable Content: Depending on the area and environment, ranges from 0.1 percent to 4 percent, and plays a significant influence in determining the weather and climate.

B. Layers of the Atmosphere

There are five primary layers that make up the atmosphere of the Earth, and each of these layers has its own unique features in terms of temperature, physical makeup, and function. The following is an outline of the layers, beginning with the surface of the Earth and working outward:

a. Troposphere: It stretches between 12 and 18 kilometers from the Earth's surface. All meteorological phenomena, including clouds, rain, and storms, take place in this stratum. In this layer, the temperature drops with height. It is the densest layer of the atmosphere and makes up around 75% of its mass. The tropopause, which divides the troposphere from the following layer, is where it terminates.

b. Stratosphere: Its length ranges from roughly 12 km to 50 km. The ozone layer, which it includes, scatters and absorbs ultraviolet sun radiation. Because ozone absorbs UV rays, the temperature rises with height in contrast to the troposphere. Commercial aircraft frequently fly in the stratosphere since there is relatively little weather or turbulence there.

c. Mesosphere: It stretches from approximately 50 km to 80-90 km. Due to the fact that the temperature in this layer drops with increasing altitude; it is the layer of the atmosphere that is the coldest. Upon entering the atmosphere, the majority of meteors are destroyed by the fire that occurs at this layer. As the mesosphere transitions into the subsequent layer, the menopauses serve as the connecting point.

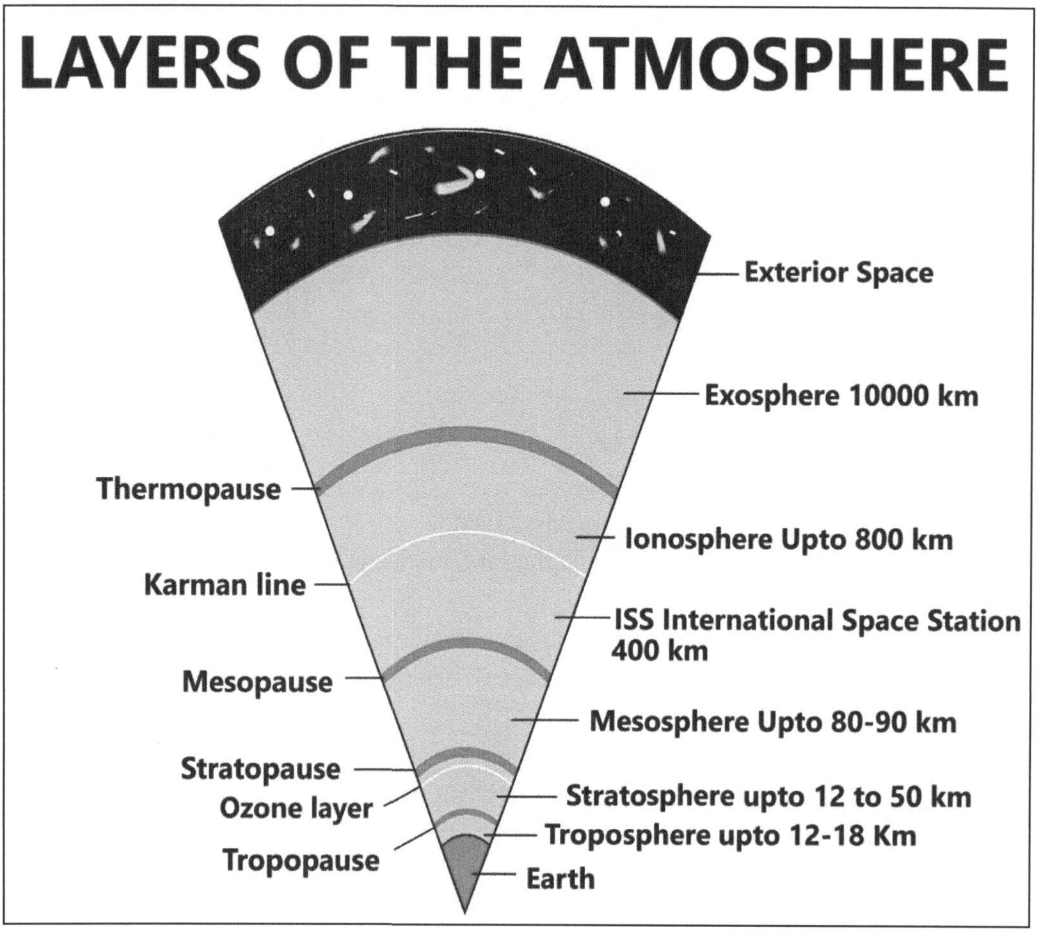

Figure 16.1: Layers of atmosphere.

d. Thermosphere: Its length ranges from roughly 88 to 90 km to 800 km. Although the thin air would not feel hot, the temperature in this layer rises with altitude and can reach 2,500°C (4,500°F) or higher. Long-distance communication is made possible by the ionosphere, a sublayer of this layer made up of ionized particles that reflect radio signals. In this layer, solar winds interact with charged particles to produce the auroras, also known as the Northern and Southern Lights.

The Kármán Line

The Kármán Line, which is situated at an elevation of approximately 100 km (62 miles) above sea level, is widely acknowledged as the border that represents the barrier between the atmosphere of Earth and the infinite space beyond it.It is named after Theodore von

Kármán, a Hungarian-American engineer and physicist, and it denotes the point at which the atmosphere becomes too thin to enable conventional aircraft flying. This is because the air density is insufficient for aerodynamic lift. After reaching this particular height, spacecraft are needed to rely on orbital velocity rather than aerodynamic forces in order to maintain motion.When it comes to defining space flight and distinguishing between astronauts and pilots, the Kármán Line is a crucial reference that is utilized by international organizations such as the Fédération AéronautiqueInternationale (FAI).Although many organizations, such as the United States military and the National Aeronautics and Space Administration (NASA), employ slightly different altitudes to define space, the Kármán Line is commonly acknowledged as the official boundary.

e. **Exosphere:** Between 800 - 10,000 km is the range that it covers. At this point, the atmosphere begins to evaporate into space, making it the outermost layer. There are only a few particles in it, and they are able to move for hundreds of kilometers without coming into contact with one another. A significant number of satellites orbit within this stratum.

Layer	Approximate Altitude	Key Features
Troposphere	0 - 15 km	Weather, clouds, temperature decreases
Stratosphere	15 - 50 km	Ozone layer, temperature increases
Mesosphere	50 - 85 km	Meteors burn up, temperature decreases
Thermosphere	85 - 600 km	High temperatures, auroras, ionosphere
Exosphere	600 km - 10,000 km	Thin, outer space-like atmosphere

Table 16.1: Layers of atmosphere and their key features

8. Human Impact on the Atmosphere

A. Industrialization and Pollution

a. Greenhouse Gases:

Increased Emissions: There has been an increase in the concentration of greenhouse gases (CO_2 and CH_2 and N_2O) due to the burning of fossil fuels, deforestation, and industrial operations, which has contributed to the phenomenon of global warming.

b. Air Pollution:

Particulate Matter and Gases: This is because the emission of pollutants like sulfur dioxide (SO_2), nitrogen oxides (NO_2), and particulate matter has an impact on both the quality of the air we breathe and the health of people.

B. Ozone Depletion

a. Chlorofluorocarbons (CFCs):

Destruction of Ozone: CFCs, when released into the atmosphere, have the ability to degrade ozone molecules, resulting in a reduction of the ozone layer and an increase in ultraviolet radiation at the surface.

b. Montreal Protocol:

Global Agreement: This international pact, which was adopted in 1987, aims to eliminate the production and utilization of compounds that deplete the ozone layer, so facilitating the gradual restoration of the ozone layer.

Activities like as volcanic eruptions, chemical reactions, biological processes, and variations in climate have all had a role in the formation and development of the atmosphere that surrounds the Earth. Some of the processes that have contributed to the formation of the atmosphere that we have today include the elimination of the primary atmosphere, the formation of a secondary atmosphere that is rich in nitrogen, and the oxygenation that is driven by creatures that are capable of photosynthesis. It is essential to have a solid understanding of these processes in order to take on the present difficulties that are associated with climate change, air pollution, and environmental sustainability.

Chapter 17
The Formation of the Biosphere

Introduction

The biosphere is comprised of all living species that exist on Earth as well as the interactions that these organisms have with their outer environment. The birth of the biosphere and its subsequent evolution are both complex processes that take place over billions of years and involve interactions between geological, chemical, and biological elements. After gaining an understanding of these processes, one can have a better understanding of the beginnings of life, the evolution of ecosystems, and the influence that life has had on the environment of Earth.

1. Origin of Life

A. Prebiotic Chemistry

a. Miller-Urey Experiment:

- Objective: Simulated early Earth conditions to investigate the formation of organic molecules.
- Findings: Under the impact of electrical sparks, which simulate lightning, it was demonstrated that simple organic compounds, such as amino acids, may be produced from inorganic ingredients such as methane, ammonia, hydrogen, and drinking water.

b. Hydrothermal Vents:

- Environment: Hydrothermal vents located deep within the ocean offer circumstances that are characterized by high temperatures, pressures, and chemical gradients.
- Role: These particular conditions have the potential to have helped the process of synthesising complex organic compounds and offered sources of energy for the earliest forms of life.

c. **Panspermia Hypothesis**:

- Concept: This theory proposes that the origin of life or its ancestors could have been brought to Earth from outer space by means of comets or meteorites.
- Evidence: Meteorites have been shown to contain organic molecules, including amino acids, which lends support to the the concept that the fundamental components of life may have originated from extraterrestrial sources.

B. Formation of Protocells

a. Lipid Bilayers:

- Self-Assembly: Amphiphilic substances, including fatty acids, have the ability to spontaneously form lipid bilayers in aqueous settings, hence resulting in the formation of primitive cell membranes.
- Protocells: It is possible for these lipid bilayers to enclose organic molecules, thereby producing protocells that are capable of performing fundamental metabolic processes.

b. **RNA World Hypothesis**:

- Self-Replicating RNA: It is possible that RNA molecules, which possess the ability to catalyze chemical reactions and retain genetic information, could have been the initial substance that possessed genetic information.
- Evidence: Ribozymes, which are enzymes derived from RNA, provide evidence that RNA is capable of carrying out catalytic actions that are essential for the survival of living organisms.

2. Evolution of Early Life

A. Prokaryotic Life

a. **First Prokaryotes**:

- Appearance: It is most probable that the initial forms of life were uncomplicated, single-celled prokaryotes, such as bacteria and archaea, which originated approximately 3.8 billion years ago.
- Anaerobic Metabolism: Anaerobic fermentation and other anaerobic processes were the primary means by which early prokaryotes generated energy despite the absence of oxygen.

b. **Stromatolites**:

- **Formation**: It is the action of photosynthetic cyanobacteria that results in the formation of layered structures.

- **Significance**: Some of the earliest fossil evidence of life can be found in stromatolites, which date back more than 3.5 billion years.

B. Oxygenic Photosynthesis

a. Cyanobacteria:

- Innovation: Through the utilization of sunlight, oxygenic photosynthesis was developed, which involved the transformation of water and carbon dioxide into organic matter and oxygen.
- Impact: Oxygen was released into the atmosphere, which resulted in the gradual oxygenation of the environment on Earth during the process.

b. Great Oxygenation Event (GOE):

- Timeline: Occurred around 2.4 billion years ago.
- Consequences: As the levels of oxygen in the atmosphere rise, it causes the oxidation of iron in the oceans, which results in the production of banded iron formations. This, in turn, leads to the formation of the ozone layer.

3. Development of Eukaryotic Cells

A. Endosymbiotic Theory

a. Concept: It has been proposed that the origin of eukaryotic cells can be traced back to a mutually beneficial connection between various bacterial populations.

b. Mitochondria and Chloroplasts: There is evidence to suggest that these organelles started from bacteria that were free-living and were subsequently consumed by eukaryotic cells that were ancestral to the organism.

- Mitochondria: Derived from proteobacteria, enabling aerobic respiration.
- Chloroplasts: Derived from cyanobacteria, enabling photosynthesis in plants and algae.

B. Evolution of Multicellularity

a. Advantages:

- Specialization: Because of multicellularity, cells were able to specialize for various functions, which led to an increase in the complexity of the organism.
- Size and Efficiency: It is possible for larger creatures that possess several cells to take advantage of new ecological niches and enhance resource usage.

b. Timeline:

- Earliest Multicellular Life: From around 1.6 billion years ago, fossils of primitive multicellular creatures have been discovered all across the world.
- Cambrian Explosion: At approximately 541 million years ago, there was a significant increase in the diversity of multicellular life forms, which ultimately resulted in the formation of the majority of the main animal phyla.

4. Colonization of Land

A. **Adaptation to Terrestrial Environments**

a. **Plants and Fungi**:

- Symbiosis: Mycorrhizae, which are fungi, are likely to have created symbiotic associations with early plants, which facilitated the uptake of nutrients.
- Adaptations: Through the development of various structures, such as roots, vascular tissues, and protective cuticles, plants were able to flourish in terrestrial environments.

b. **Animals**:

- Transition from Water to Land: As a result of their evolution from fish, early amphibians developed adaptations such as lungs and limbs to thrive in terrestrial environments.
- Diversification: Insects, reptiles, birds, and mammals all underwent subsequent evolution, with each species individually adapting to a distinct terrestrial habitat.

B. **Formation of Soil and Ecosystems**

a. **Soil Formation**:

- Biological Activity: One of the factors that led to the development of soil was the decomposition of organic matter by microbes, as well as the physical and chemical weathering of rocks.
- Nutrient Cycling: The establishment of nutrient cycles, which are crucial for the maintenance of terrestrial ecosystems, is a result of planting and soil organisms.

b. **Ecosystem Development**:

- Biodiversity: The formation of distinct ecosystems was a result of the interaction between various creatures, including plants, animals, fungus, and bacteria.
- Energy Flow and Food Webs: Primary producers, which include plants and algae, were responsible for capturing solar energy, which in turn supported food webs that comprised herbivores, carnivores, and decomposers.

5. Modern Biosphere

A. **Components of the Biosphere**

a. **Ecosystems**: The communities of living organisms that are interacting with the physical environment in which they are found. Both terrestrial ecosystems (such as forests, grasslands, and deserts) and aquatic ecosystems (such as oceans, rivers, and lakes) are included in this category.

b. **Biomes**: Large areas that are distinguished by certain climatic conditions and the types of vegetation that predominate in those areas. Deserts, tundra, taiga, temperate woods, tropical rainforests, and savannas are some examples of this type of environment.

B. **Interactions within the Biosphere**

a. Biogeochemical Cycles:

- Carbon Cycle: The exchange of carbon between the atmosphere, hydrosphere, lithosphere, and biosphere is a process that is driven by several processes such as photosynthesis, respiration, and decomposition.
- Nitrogen Cycle: Comprises the processes of nitrogen fixation, nitrification, assimilation, and denitrification, all of which are essential for the synthesis of nucleic acids and proteins.

b. Energy Flow:

- Trophic Levels: The flow of energy includes the movement of energy from primary producers to different tiers of consumers, including herbivores, carnivores, and decomposers.
- Efficiency: The transmission of energy across different trophic levels is often inefficient, resulting in a large loss of energy in the form of heat.

6. Human Impact on the Biosphere

A. Environmental Changes

a. Deforestation:

- Impact: This results in the destruction of habitats, a decrease in biodiversity, and changes in temperature as well as the storage of carbon.
- Causes: Driven by agricultural expansion, logging, and urbanization.

b. Pollution:

- Types: It encompasses air pollution, which is caused by emissions from industrial operations, water pollution, which is caused by chemical runoff and plastic debris, and soil contamination, which is caused by pesticides and heavy metals.
- Consequences: There is a significant impact on ecosystems, human health, and climate change as a result of this.

B. Climate Change

a. Global Warming:

- Causes: There has been an increase in the concentrations of greenhouse gases (CO_2, CH_2, and N_2O) due to human activities such as the burning of fossil fuels and the destruction of forests.
- Effects: There is an increase in temperatures, a shift in the patterns of precipitation, the melting of ice caps, and an increase in sea levels.

b. Impact on Ecosystems:

- Habitat Shifts: Due to the fact that their typical habitats are becoming unsuitable, species are being compelled to relocate to other locations.
- Extinctions: There are certain species that might not be able to adapt well, which could result in higher rates of extinction.

For billions of years, the origin and evolution of the biosphere are the consequence of complex interactions between geological, chemical, and biological processes. These interactions have been happening simultaneously. The biosphere has been subjected to a multitude of transformations, beginning with the emergence of life in the primordial oceans and continuing through the colonization of land and the development of a wide variety of ecosystems. It is essential to have a profound understanding of these processes in order to effectively handle the modern concerns that are associated with biodiversity, environmental sustainability, and climate change.

Chapter 18
The Age of the Earth

Introduction

Throughout the history of geology, astronomy, and other scientific disciplines, one of the most important questions has centered about how old the Earth is. The precise age of the Earth has been determined through the application of modern technology, which has thrown light on the processes that have shaped our planet over the course of billions of years.

1. **Historical Perspectives**

A. **Ancient Religious and Mythological Theories:** A great number of historically significant societies had their very unique origin myths, which frequently suggested that the Earth was only a few thousand years old.

B. **Early Scientific Approaches**

a. James Ussher (1650): According to the genealogy found in the Bible, an Irish Archbishop was the one who determined that the Earth was formed in the year 4004 BCE.

b. Georges-Louis Leclerc, Comte de Buffon (1779): The age of the Earth was estimated to be at least 75,000 years old by comparing it to the rate at which iron spheres cool throughout the cooling process.

2. **Development of Modern Geological Methods**

A. **Uniformitarianism**

a. James Hutton (late 18th century): In order to create the groundwork for the theory of uniformitarianism, it was proposed that the features of the Earth were formed by processes that occurred during enormous periods of time.

b. Charles Lyell (1830-1833): Hutton's ideas were expanded upon in his work "Principles of Geology," in which he advocated for progressive geological processes and suggested that the Earth in question was an ancient planet.

B. **Stratigraphy and Fossils:** At the beginning of the 19th century, B. William Smith created the idea of faunal succession and utilized fossil correlation in order to determine the relative ages of rock layers. One of the contributions that William Smith made to the process of relative dating geological formations was the discovery that particular fossils were regularly discovered in particular strata.

3. **Radiometric Dating Techniques**

A. **Principles of Radiometric Dating:** Decomposition of unstable isotopes into stable isotopes occurs at a rate that can be predicted and is referred to as the half-life. The age of a sample can be determined by measuring the ratio of parent to daughter isotopes in the sample.

B. **Key Radiometric Methods**

a. **Uranium-Lead Dating**: Utilises the process by which Uranium-238 and Uranium-235 decay into Lead-206 and Lead-207 respectively. This technique is useful for dating rocks that are more than one million years old, and it has also been utilized to date the minerals that on Earth are the oldest.

b. **Potassium-Argon Dating**: The process of converting potassium-40 into argon-40 is utilized. There is a range of millions to billions of years that may be determined using this method, which is useful for dating volcanic rocks.

c. **Rubidium-Strontium Dating**: Strontium-87 is produced through the process of rubidium-87 decaying. Over a wide variety of geological periods, this technique can be utilized to date rocks and minerals based on their composition.

d. **Carbon-14 Dating**: Nitrogen-14 is produced by the process of carbon-14 decay. The Earth itself cannot be dated using this approach; however it is effective for dating organic items that are up to around 50,000 years old.

4. **Determining the Age of the Earth**

A. **Early Estimates**

a. Clarence King (1893): On the basis of the rate at which the Earth is cooling, it has been estimated that the age of the Earth is one hundred million years.

b. John Joly (1899): On the basis of the accumulation of salt in the oceans, it was estimated that the age of the Earth fell somewhere between 90 and 100 million years.

B. Radiometric Age Determinations: Arthur Holmes (1911-1956): The age of the Earth was initially determined to be 1.6 billion years by an early researcher who utilized radiometric dating to assess the age of rocks. This scientist was the first to make this estimation.

C. The Age of Meteorites: The origin of the solar system can be dated using meteorites, which are thought to be remains of the early solar system. Meteorites also provide insight into the age of the solar system. By applying the technique of Uranium-Lead dating on meteorites in the 1950s, Clair Patterson was able to ascertain that the age of the Earth is around 4.55 billion years.

5. Current Understanding

A. Modern Estimates: There is a level of uncertainty of roughly one percent regarding the age of the Earth, which is generally accepted to be approximately 4.54 billion years.

B. Supporting Evidence

a. Oldest Terrestrial Rocks: There is a rock formation in Canada known as the Acasta Gneiss that has been dated to around 4.0 billion years. It has been determined that the age of zircon minerals found in the Jack Hills in Australia is approximately 4.4 billion years.

b. Lunar Samples: Some of the samples collected during the Apollo mission have revealed ages that are in agreement with the origin of the Earth-Moon system approximately 4.5 billion years ago.

c. Helioseismic and Stellar Models: It is estimated that the Sun and the solar system have been around for around 4.6 billion years, according on models of solar and stellar evolution.

6. Implications of Earth's Age

A. Geological Time Scale: To begin, the geological time scale is broken down into eons, which include the Hadean, Archean, Proterozoic, and Phanerozoic periods. These eons are then further separated into eras, periods, and epoch. Each division is a reflection of key events that have occurred throughout the history of the Earth, such as the arrival of life, the development of continents, and the occurrence of mass extinctions.

B. Biological Evolutionary: Because of the vast amount of time that has passed since the beginning of Earth's history, life has been able to gradually evolve from simple microbes to sophisticated multicellular organisms. The Cambrian Explosion, the rise of vertebrates, the colonization of land, and the evolutionary development of animals and humans are all significant events that occurred during this time period.

C. Planetary Development

a. Formation and Differentiation: The processes of planetary formation, differentiation into core, mantle, and crust, and the creation of a stable atmosphere and hydrosphere can be better understood with a better understanding of the age of the Earth in order to gain a better understanding of these processes.

b. Tectonic Activity: In the first place, the age of the Earth makes it possible to investigate the processes of plate tectonics, mountain development, and the formation of geological features over extended periods of time.

In geological research, one of the most important concepts is the age of the Earth, which has been estimated to be roughly 4.54 billion years. Our understanding of the history and dynamics of the Earth has been completely transformed as a result of the discovery of radiometric dating techniques. With this information, a framework may be established for the study of the evolution of the planet, the development of life, and the dynamic processes that continue to shape the Earth in the present day.

Chapter 19
Continental Drift Theory

Introduction: In 1912, Alfred Wegener, a German geophysicist and meteorologist, put out the idea that Continental Drift Theory should be considered. According to this hypothesis, continents have moved around throughout the course of geological time and were previously united to form a single supercontinent altogether.

Key Concepts

1. **Pangaea**: Wegener put out the idea that all of the continents comprised a massive landmass known as Pangaea at one point in time. One of the supercontinents that existed during the late Paleozoic and early Mesozoic eras is known as Pangea.The theory put forth by Wegner suggests that the disintegration of Pangaea started approximately 200 million years ago, during the Jurassic period.
2. **Movement of Continents:** Over the course of millions of years, continents gradually moved out of their original places. The concept that continents are not fixed but rather float on the surface of the Earth was the foundation upon which the theory was built.As a result of continents breaking through the more solid oceanic crust, this movement was comparable to that of an iceberg moving through water.

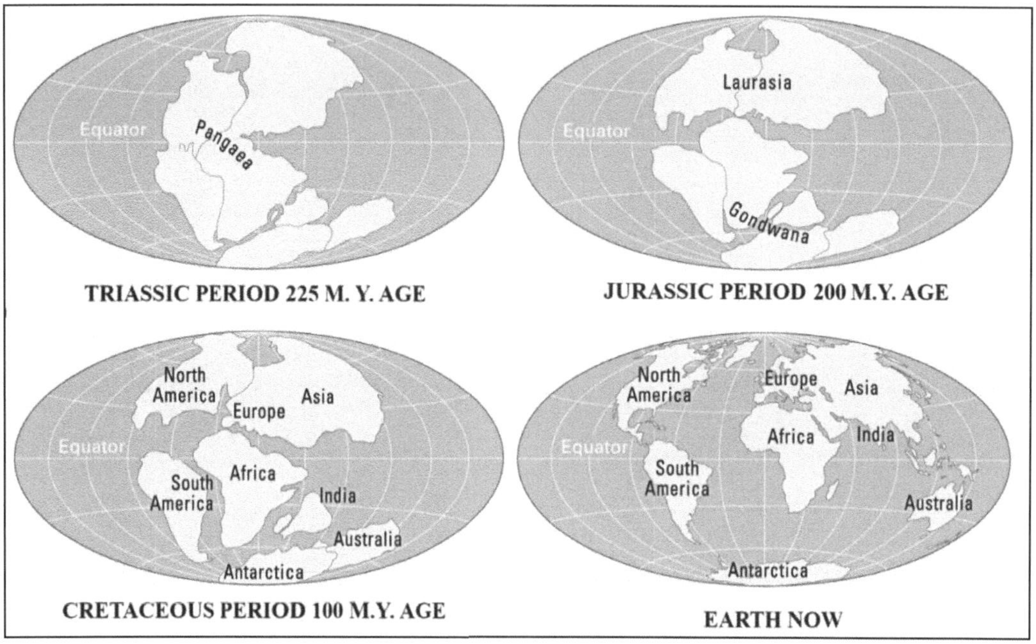

Figure 19.1: Illustration of Continental Drift Theory

Evidence Supporting Continental Drift

1. **Fit of the Continents**: The fact that the continents fit together like a jigsaw puzzle is one of the most compelling pieces of evidence that supports the theory of continental drift. South America and Africa, in instance, have coastlines that appear to be precisely aligned with one another, which suggests that they were previously united together.

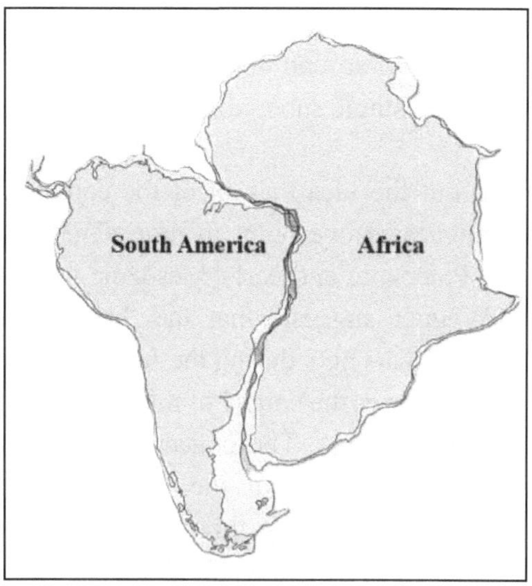

Figure 19.2: Figure illustration Fit of the Continents

2. **Fossil Evidence**: Fossils of plants and animals that are identical to one another have been discovered on continents that are now separated by large oceans. Take, for example:

A. **Mesosaurus**: A reptile that lives in watery environments and whose fossils have been discovered in both South America (Brazil) and Southern Africa for example. Given that Mesosaurus was a creature that lived in freshwater, it is highly improbable that it could have traveled across the Atlantic Ocean. This suggests that these landmasses were formerly connected to one another.

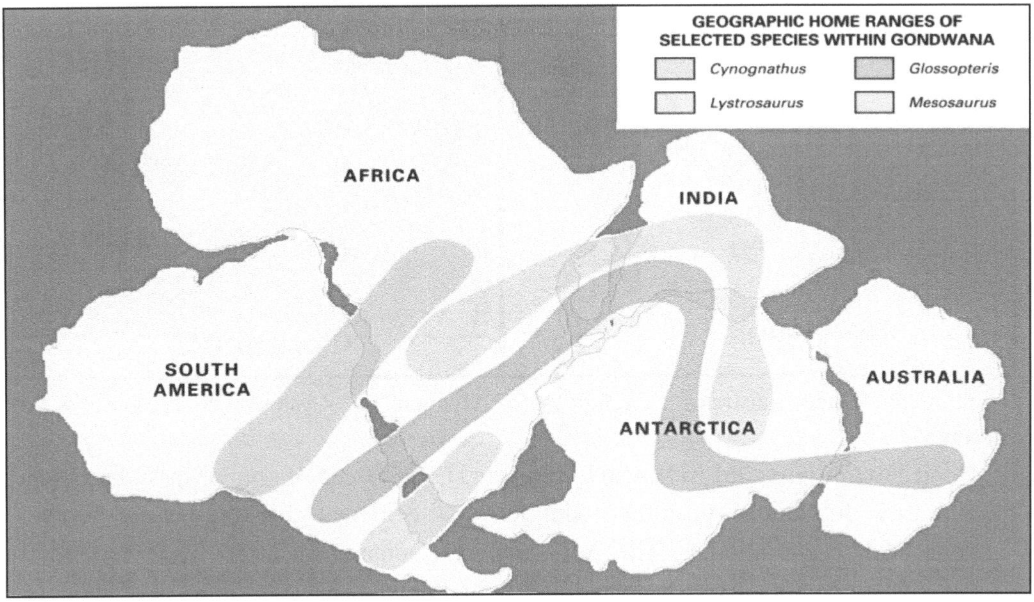

Figure 19.3: Figure illustrates Fossil Evidence from past (Reference: National Geographic Society. "Continental Drift)

B. **Glossopteris**: This particular genus of seed fern has been found on a number of different continents, including South America, Africa, Antarctica, Australia, and India. The vast distribution of this species provides evidence that these regions were originally a part of a single landmass known as Gondwana.

C. **Lystrosaurus**: A herbivorous reptile that can be discovered in South Africa, Afghanistan, and Antarctica. Indications that Lystrosaurus fossils can be found on a number of different continents hint to a period of time when these continents were connected together as a component of the supercontinent Pangaea.

D. **Cynognathus**: The fact that remains of this land-dwelling predator have been discovered in both South Africa and South America lends credence to the theory that these two continents were formerly next to one another.

Figure 19.4: Figure illustrates morphology of Cynognthus, Lystosaurus, mesosaurus and glossopteris.

3. **Rock Formations and Mountain Ranges**: The presence of similar mountain ranges and rock formations on other continents is evidence that they were formerly connected to one another. The Appalachian Mountains in North America and the Caledonian Mountains in Scotland and Scandinavia are two examples of mountain ranges that you can encounter. Rock layers that are comparable can be found in South Africa, Brazil, and Argentina.

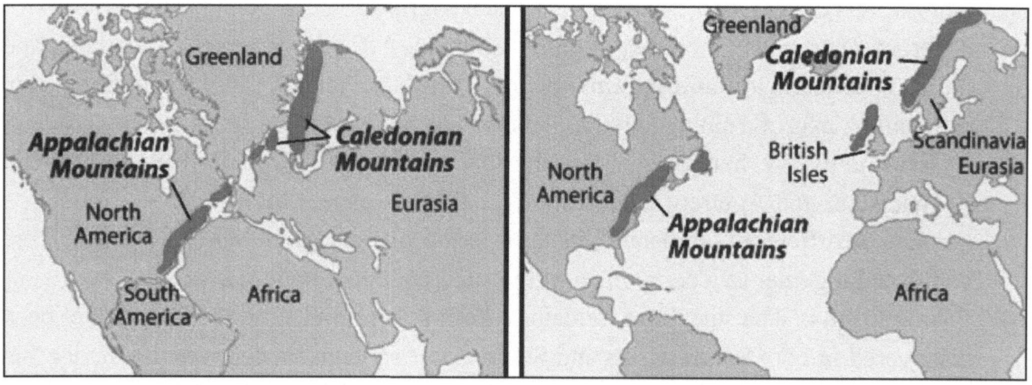

Figure 19.5: Figure illustrates before and after continental drift of Appalachian Mountains in North America and the Caledonian Mountains in Scotland and Scandinavia

4. **Glacial Deposits in Tropical Regions:** Geologists have discovered evidence of glaciation in present tropical countries, such as sections of Africa, South America, India, and Australia. This evidence can be found in the form of glacial patterns and tillites, which are sedimentary deposits that glaciers have left behind. Based on these characteristics, it appears that these continents were formerly situated closer to the poles, which would have been more suitable for glacial. Glacial deposits dating back to the late Paleozoic period, which occurred approximately 300 million years ago, have been discovered in these places. This finding provides evidence that these regions were formerly a part of the supercontinent known as Gondwana, which was situated close to the South Pole.
5. **Paleoclimate Evidence**: The discovery of coal in Antarctica, which is formed by the buildup of plant material in conditions that are warm and swampy, brought about the discovery of coal. This provides evidence that the continent was originally situated closer to the equator, where a tropical environment would have been advantageous for the development of the luxuriant flora that is required for the creation of coal. The movement of Antarctica from a tropical zone to its current polar position can be explained by the phenomenon known as continental drift.
6. **Mechanism of Drift**: The idea that continents moved through the oceanic crust and plowed through the sea floor was proposed by Wegener. Wegener was unable to provide an explanation for the force that was driving the drift, which led to criticism of this idea.

Rejection and Acceptance

A. Initial Skepticism and Rejection

Wegener's theory faced significant opposition from the geological community. Key reasons for this included:

- When it comes to continental movement, there is a lack of a plausible mechanism.
- There is a prevalent assumption that continents and ocean basins are permanent. More specifically, the majority of individuals believe that ocean basins and continents are forever.
- There was a limited grasp of the geological processes and the innards of the Earth throughout that historical period.

B. Revival and Acceptance

Interest in continental drift was revived in the mid-20th century with the discovery of seafloor spreading and advancements in the study of plate tectonics. Key contributions included:

- **Harry Hess** proposed the theory of seafloor spreading, suggesting new oceanic crust forms at mid-ocean ridges and spreads outward.

- **Marie Tharp** created detailed maps of the ocean floor, revealing the presence of mid-ocean ridges and supporting the idea of seafloor spreading.

Impact and Legacy

- Wegener's Continental Drift Theory revolutionized geology and led to significant advancements in understanding Earth's geological processes.
- It paved the way for modern geology and the study of Earth's dynamic nature.

Key Figures

- Alfred Wegener: Meteorologist and geophysicist, primary proponent of Continental Drift.
- Harry Hess: Proposed seafloor spreading, contributing to Plate Tectonics Theory.
- Marie Tharp: Created detailed maps of the ocean floor, supporting seafloor spreading.

Chapter 20
Sea Floor Spreading

There are fundamental principles in geology that describe the dynamic character of the lithosphere of the Earth. One of these concepts is sea floor spreading. The mechanics of sea floor spreading, the evidence that supports it, and the vital role that it plays in the theory of plate tectonics are all taken into consideration in this topic. In order to understand the production and movement of oceanic crust, the distribution of geological activity, and the evolution of the Earth's surface, it is essential to have a solid understanding of these processes.

Definition and Mechanisms

Sea floor spreading is the process by which new oceanic crust is generated at mid-ocean ridges and then progressively spreads away from the ridge, to create new sea floor. This process is known as sea floor spreading. Harry Hess was the first person to suggest this process in the early 1960s. It served as a mechanism that offered support for Alfred Wegener's previous hypothesis of continental drift.

Mechanisms of Sea Floor Spreading

1. **Initial Rifting Stage**

A. **Tensional Forces:** In the beginning of the process, tensional forces in the lithosphere of the Earth begin to stretch and thin the continental crust. In most cases, these forces are brought about by convection currents deep below the mantle, which is located beneath the lithosphere.

B. Faulting: As a result of the stretching of the crust, it breaks along normal faults, which results in the formation of rift valleys. The terrain is characterized by the formation of grabens, which are blocks that have been dropped down, and horsts, which are blocks that have been hoisted up.

C. Volcanic Activity: As the crust undergoes a process of thinning, magma emerging from the mantle starts rising to the surface by way of the fractures contained within the thinning crust. Additionally, this has the potential to cause volcanic activity throughout the rift zone.

D. Example: The East African Rift Valley is a modern example of this early rifting stage, where large cracks in the Earth's surface have formed as the African Plate is being pulled apart.

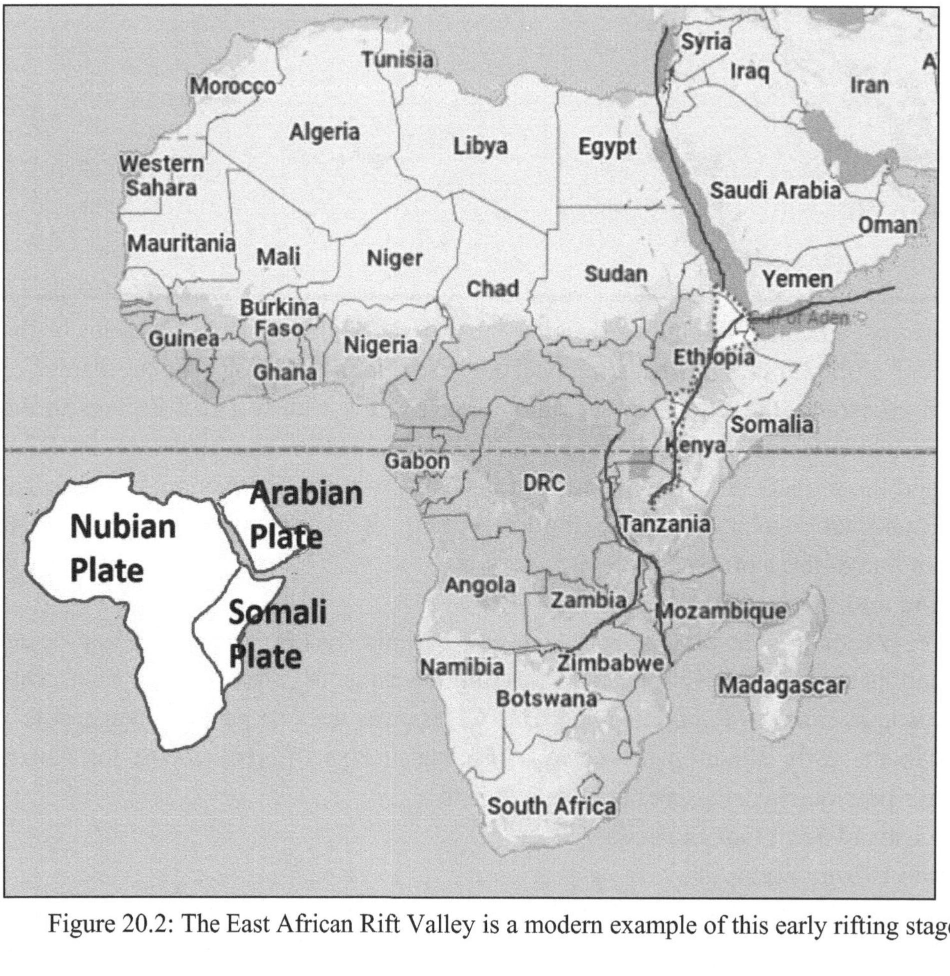

Figure 20.2: The East African Rift Valley is a modern example of this early rifting stage.

2. **Formation of Rift Valleys**

A. **Continued Extension:** The process of rifting continues, causing the lithosphere to stretch further, which results in deeper faulting and the construction of a rift valley that is more evident. In the middle of the rift, the crust takes on a substantially thinner appearance.

B. **Subsidence:** An area of topographical lowness is created as a result of the core rift valley collapsing. This region may eventually get filled with water, resulting in the formation of lakes or shallow seas as it sinks.

C. **Volcanism Increases:** As the crust continues to weaken, magma continues to rise through it, and volcanic activity becomes more prominent. This activity frequently results in the formation of shield volcanoes or fissure eruptions along the rift.

D. **Example:** The Red Sea is in this stage, where a rift valley is forming between the African and Arabian plates and is gradually filling with water.

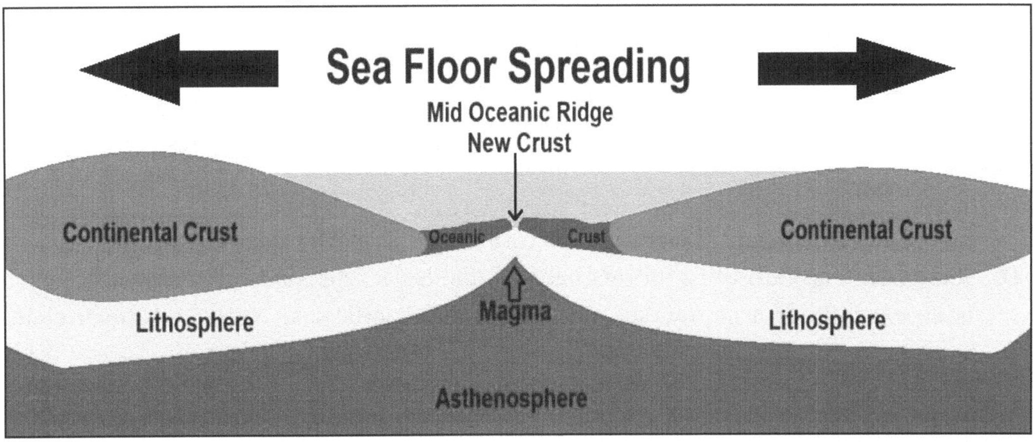

Figure 20.1: Diagram of Sea Floor Spreading: Illustrating the process of new crust formation at mid-ocean ridges

3. **Formation of a Narrow Sea**

A. **Breakup of Continental Lithosphere:** At some point in time, the process of rifting ultimately results in the full disintegration of the continental lithosphere. It is possible that the center fissure will drop to the point where it will allow water from the ocean to pour in, resulting in the formation of a narrow sea.

B. **Oceanic Crust Formation Begins:** The continental crust is continuously stretching and thinning, causing magma to rise from the mantle. This magma then cools and solidifies, resulting in the formation of new oceanic crust. The beginning of the spreading of the sea floor can be seen here.

C. **Narrow Ocean Basin:** When the rift valley reaches this stage, it undergoes a transformation into a narrow ocean basin, which is comparable to a newborn ocean.

At the same time that the two continental blocks continue to move apart, the process of sea-floor spreading results in the formation of new oceanic crust.

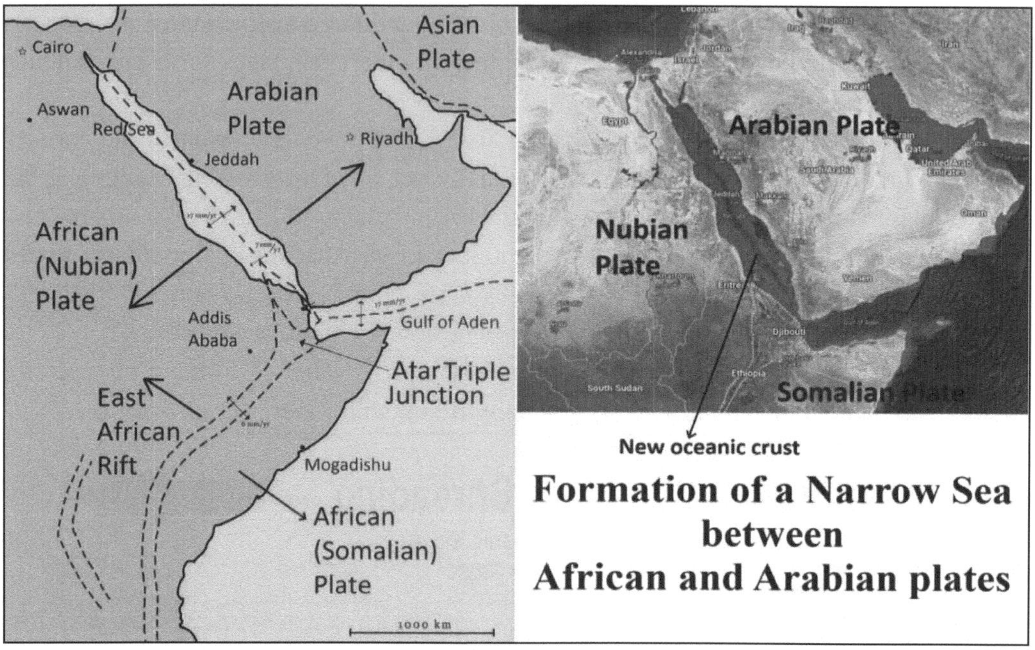

Figure 20.3: Figure illustrating the Formation of Narrow Sea from a Rift Valley

D. Example: The Gulf of California between the Baja Peninsula and mainland Mexico is an example of a narrow sea forming as the Pacific and North American plates spread apart.

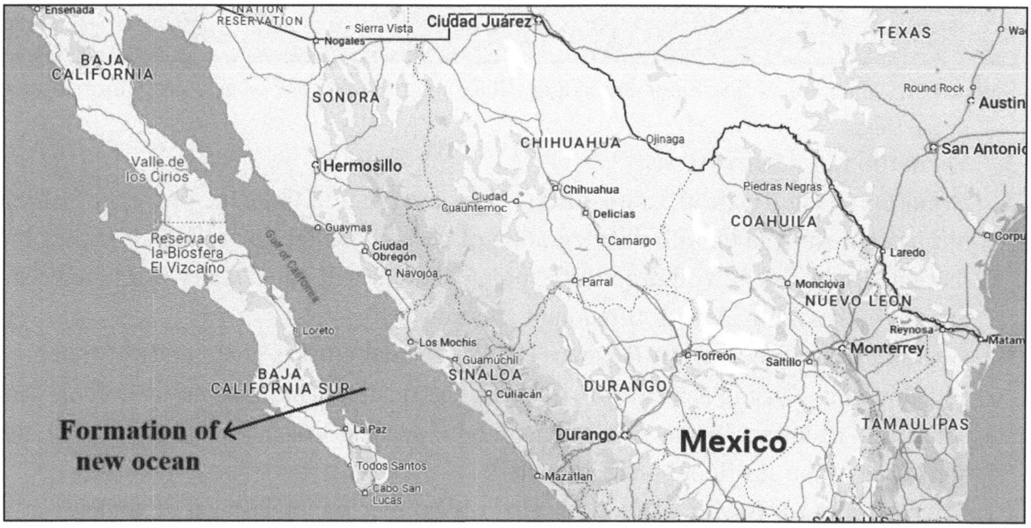

Figure 20.4: Formation of a narrow sea between the Baja Peninsula and mainland Mexico main land.

4. **Mature Ocean Basin**

A. **Full Ocean Basin and Mid-Ocean Ridge Formation:** The process of rifting continues, resulting in the formation of a fully developed ocean basin. A mid-ocean ridge that runs in the center of the new ocean is the only thing that separates the two portions of the old continent, which are now separated by thousands of kilometers.

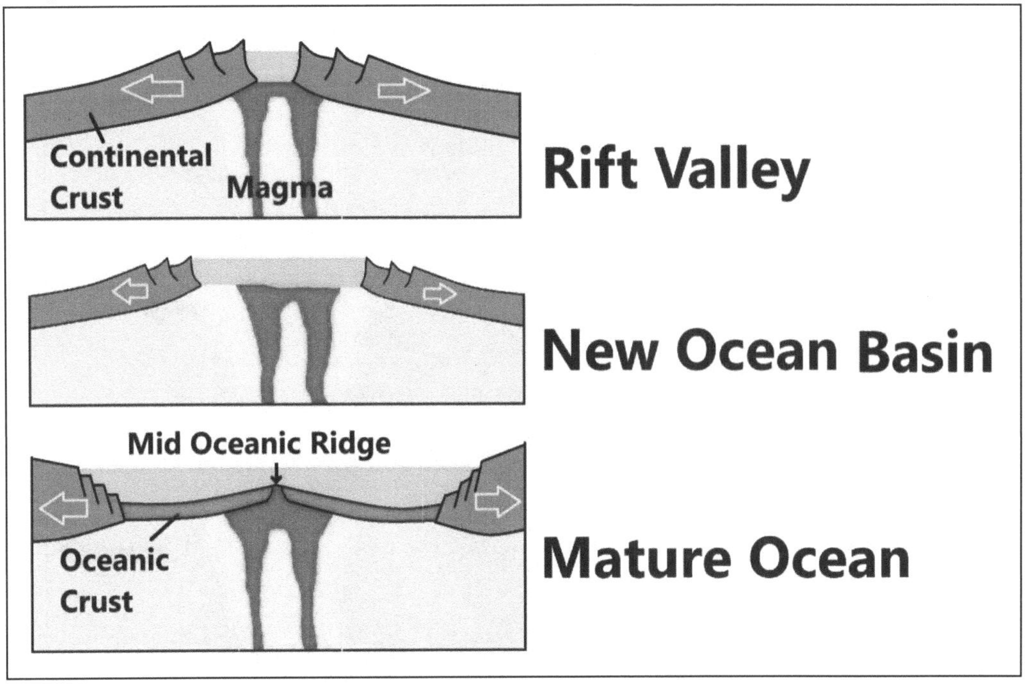

Figure 20.5: Diagram of Stage of Sea Floor Spreading.

B. **Mid-Ocean Ridge Spreading:** The ongoing movement of magma along the mid-ocean ridge results in the formation of new oceanic crust, which in turn causes the plates to move apart. As additional crust is produced, the ocean basin expands to accommodate it.

C. **Symmetric Sea-Floor Spreading:** Over time, the ocean basin widens as a result of the symmetrical spreading of oceanic crust on either side of the mid-ocean ridge.

D. **Example:** The Atlantic Ocean, which formed as the America drifted away from Europe and Africa, is a mature ocean basin where sea-floor spreading continues along the Mid-Atlantic Ridge.

Evidences of Sea Floor Spreading

This theory is supported by multiple lines of data, offering a strong foundation for comprehending the dynamic character of the Earth's lithosphere.

1. **Magnetic Anomalies**

A. **Symmetrical Magnetic Stripes**: On either side of mid-ocean ridges, magnetic measurements of the ocean floor show symmetrical patterns of magnetic stripes. As fresh oceanic crust emerges and shifts away from the ridge, these stripes show times of normal and reversed magnetic polarity. The assumption that fresh crust is constantly being created and pushed outward is supported by the stripes' symmetry on each side of the ridge axis.

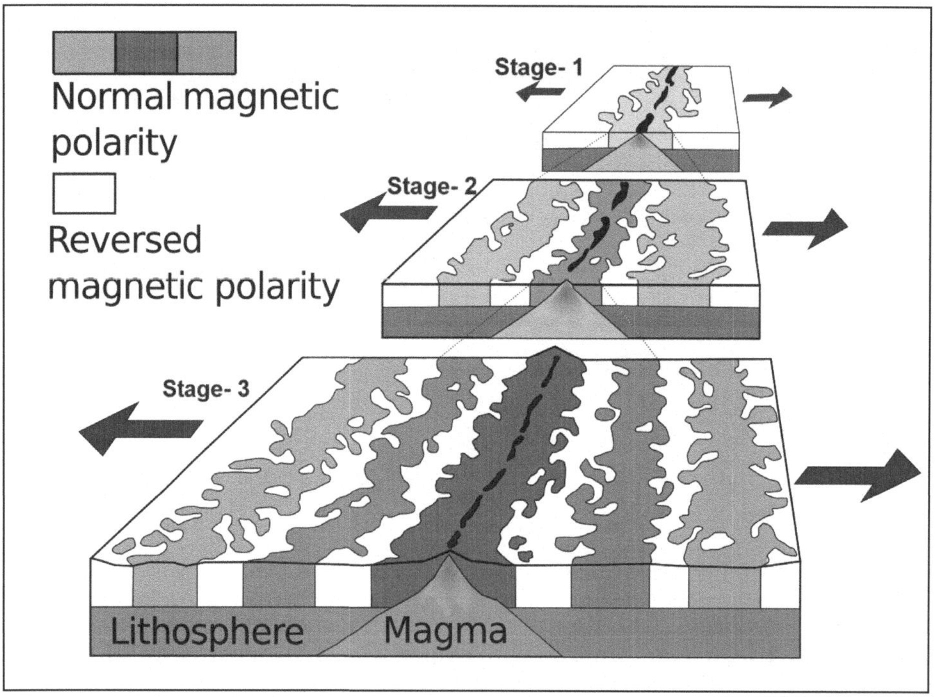

Figure 20.6: Formation of Magnetic anomalies at Mid Oceanic ridge. (Reference: "Plate Tectonics."Wikipedia, The Free Encyclopedia)

B. **Magnetic Reversals**: Over the course of geological history, the magnetic field of the Earth has undergone numerous instances of inversion. The magma that is solidifying at the mid-ocean ridges aligns with the magnetic field that is now present, thereby maintaining a record of the direction in which the magnetic field is moving. A pattern of magnetic stripes is created by this record, which mirrors each other on opposite sides of the ridge. This record provides compelling evidence for the spreading of the sea floor.

2. **Age of Oceanic Crust**

A. **Radiometric Dating**: One method that has been utilized for determining the age of oceanic crust samples is the utilization of radiometric dating techniques, namely potassium-argon dating.The findings indicate that the age of the crust has been shown to grow as the distance from the mid-ocean ridges increases. The rocks that are the youngest in age are found at the ridge axis.The ongoing production of new crust at the ridges and its subsequent outward movement is consistent with this age gradient, which is consistent with geological processes.

B. **Sediment Thickness**: As one moves further away from the mid-ocean ridges, the thickness of the sediment that covers the oceanic crust experiences a rise.On the other hand, younger crust along the ridges has relatively thin sediment layers, whereas older crust has had more time to acquire sediment during its formation.This particular pattern of sediment deposition is in accordance with the process of spreading the sea floor that occurs.

3. **Heat Flow Measurements**:Measurements of heat flow reveal that there is a larger flow of heat along mid-ocean ridges, which are the locations where new crust is being created. The assumption that mid-ocean ridges are locations of active volcanic activity and crust creation is supported by the distribution of heat flow and its associated characteristics.

4. **Seismic Activity (Earthquake Distribution)**:In the region where tectonic plates are moving apart or sliding past each other, earthquakes tend to be concentrated along mid-ocean ridges and transform faults. There is a correlation between the shallow concentration of these earthquakes and the tectonic activity that is connected with the spreading of the sea floor. There is a correlation between the pattern of seismic activity and the borders where new oceanic crust is being created and transferred.

5. **Topography of the Ocean Floor**

A. **Mid-Ocean Ridges**: The mid-ocean ridges are mountain ranges that are continuously located underwater and are generated by the upwelling of magma at the boundary of divergent plates.A central rift valley, intense volcanic activity, and hydrothermal vents are the distinguishing features of these ridges. The presence of these ridges and the traits that are connected with them offer direct evidence that provides evidence for the formation of new oceanic crust.

B. **Abyssal Plains**: Areas of the ocean floor that are flat and deep, extending away from mid-ocean ridges, are known as abyssal plains.As the oceanic crust travels away from the ridge and absorbs sediment, it undergoes a process of spreading and subsidence, which results in the formation of these structures.The shift from mid-ocean ridges to abyssal plains provides support for the notion of sea floor spreading.

6. **Geophysical Data**

A. **Gravity Anomalies**: There are differences in the gravitational field of the Earth around mid-ocean ridges and other tectonic features, which can be observed using gravity measurements.The presence of positive gravity anomalies is highly correlated with the production of dense, newly formed crust during the process of oceanic crust formation and outward movement.

B. **Seismic Reflection and Refraction**: In seismic surveys, sound waves are utilized to create an image of the structure of the marine crust and the mantle beneath it.In order to uncover the layered structure of the oceanic crust and the existence of magma chambers beneath the mid-ocean ridges, these techniques unveil the layered structure.There is evidence from seismic data that supports the concept that new crust is formed at mid-ocean ridges and then expands outward.

7. **Biological Evidence (Hydrothermal Vent Communities)**: The hydrothermal vents, which are situated along the mid-ocean ridges, enable the existence of distinct biological groups that are dependent on chemosynthesis rather than photosynthesis. These communities offer indirect evidence of volcanic and hydrothermal activity that is related with the spreading of the sea floor by providing evidence. In accordance with the process of new crust creation, the distribution of hydrothermal vents along mid-ocean ridges presents a notable alignment.

Multiple lines of evidence lend support to the idea that the sea floor is expanding. These lines of evidence include magnetic anomalies, the age and thickness of the oceanic crust, heat flow studies, seismic activity, ocean floor topography, geophysical data, and biological evidence. A thorough understanding of the dynamic processes that change the lithosphere of the Earth and contribute to plate tectonics is provided by these several sources of data, which collectively provide support for the notion of sea floor spreading.

Chapter 21
Plate Tectonics

Introduction

The development of the idea of plate tectonics in the middle of the 20th century brought about a revolutionary change in our understanding of the geological processes that occur on Earth. This theory provides an explanation for the large-scale movement of the lithosphere of the Earth, which is composed of a number of rigid plates that float on top of the asthenosphere, which is a slightly fluid layer. The underlying concepts, evidence, mechanisms, and implications of plate tectonics are investigated in this topic.

Composition of Tectonic Plates

The crust and the topmost portion of the mantle are both components of the lithosphere, which is the component that makes up tectonic plates themselves. There are many plates that make up the lithosphere, and these plates float on top of the semi-fluid asthenosphere that lies deeper below them. The geological structures and processes that are observed on Earth are influenced by these plates, which vary in size, composition, and behavior based on their characteristics. When it comes to the composition of tectonic plates, the elements that make up the lithosphere of the Earth, which includes both the crust and the top section of the mantle, are the primary factors that determine such composition. Tectonic plates, which can be either oceanic or continental in nature, or a combination of the two, organize the lithosphere into distinct layers. There is a wide range of variation among these plates with regard to their density, thickness, and the kinds of rocks that comprise them. An explanation of the components that make up tectonic plates is provided below.

1. **Oceanic Crust:** The portion of the tectonic plates that make up the oceanic crust is the section that is located beneath the ocean basins. In comparison to continental crust, it is denser, lower in thickness, and consists of a variety of rock types.

Composition:

- **Basalt**: Basalt, which is a very black and fine-grained volcanic rock, is the major type of rock that is found in the oceanic crust. It is formed when magma from the mantle rises to the surface of the ocean and cools at mid-ocean ridges.
- **Gabbro**: In addition to the basaltic layer, the oceanic crust also consists of gabbro, which is a coarse-grained volcanic rock that is comparable to basalt in composition but was generated deeper within the Earth's crust.
- **Sediments**: On the surface of the oceanic crust, there are layers of marine sediments that are very thin. These sediments include clay, silt, and organic material. When compared to the thickness of the crust itself, these sediments can exist in relatively thin layers.

Thickness: Oceanic crust is typically **5-10 km thick**.

Density: Due to the fact that oceanic crust has a higher average density of 3.0 g/cm^3, it has a tendency to sink beneath continental crust in subduction zones. This indicates that oceanic crust is denser than continental crust.

Key Characteristics: Oceanic plates are often younger than continental plates due to the fact that they are continuously formed at mid-ocean ridges and then recycled back into the mantle at subduction zones.It is mostly formed of mafic rocks, which are abundant in magnesium and iron, which are the primary contributors to the high density of the oceanic crust.

2. **Continental Crust:** The continental crust is what gives rise to the landmasses and extends beneath certain areas of the ocean that are rather shallow. When compared to oceanic crust, it is both, denser and thicker.

Composition:

- **Granite**: Granite, a light-colored, coarse-grained igneous rock, is extensively present in the continental crust. It is abundant in silica (SiO_2), potassium, and aluminum. Granite is the main component of the continental crust.
- **Metamorphic Rocks**: Metamorphic rocks, which include schist and gneiss, constitute a substantial section of the continental crust. These rocks have undergone transformations under the influence of heat and pressure within the Earth, resulting in their development.

- **Sedimentary Rocks**: The highest layers of the continental crust frequently consist of dense accumulations of sedimentary rocks such as limestone, sandstone, and shale. These rocks are formed as a result of the compression of sediments that have been around for a much longer period of time.

Thickness: A continental crust is typically between 30 to 70 kilometers thick, which is significantly thicker than oceanic crust. The thickest pieces of continental crust can be found beneath mountain ranges such as the Himalayas.

Density: The average density of the continental crust is 2.7 grams per cubic centimeter, which is lower than the density of the oceanic crust. Its reduced density enables it to "float" higher on the mantle, which ultimately results in the formation of the continents.

Key Characteristics:

In comparison to the oceanic crust, the continental crust is far older, with certain regions dating back more than 4 billion years.As a result of its composition, which is predominantly formed of felsic rocks, which are abundant in silicon, oxygen, aluminum, and other lighter elements, it is less dense than the crust of the ocean.

3. **Lithospheric Mantle:** The highest part of the mantle, which is known as the lithospheric mantle and is solid and hard, is present in both the oceanic and continental plates. The crust is the topmost layer, and this layer is a component of the tectonic plates of the earth.

Composition:

- **Peridotite**: Peridotite, a thick, coarse-grained igneous rock, is the predominant form of rock found in the lithospheric mantle. It is abundant in olivine and pyroxene, which are minerals that are rich in magnesium and iron.
- **Dunite**: There is a certain sort of peridotite that is composed of olivine in practically completely. It is common to discover this rock at a deeper level within the mantle.

Thickness: The lithospheric mantle can extend to depths of **100-200 km**.

Density: The lithospheric mantle is denser than both oceanic and continental crust, with a density of around **3.3 g/cm^3**.

Key Characteristics: As a component of the tectonic plate, the lithospheric mantle is characterized by its solidity and rigidity, which enables it to move in conjunction with the crust that lies above it.Because of its greater ductility, the asthenosphere is located beneath the mantle of the lithospheric layer, which enables the plates to move at a slower rate.

Types of Tectonic Plates

Tectonic plates are generally categorized based on the type of crust they carry: oceanic or continental. Some plates contain both types of crust.

Major Tectonic Plates

1. **Pacific Plate**: The largest tectonic plate, which is predominantly oceanic in nature. The Ring of Fire, a region with a high intensity of seismic and volcanic activity, is located on its boundary.
2. **North American Plate**: It is composed of both continental and oceanic crust on its surface. There are portions of the Atlantic Ocean that are included, and it stretches all the way to the western United States.
3. **Eurasian Plate**: It is primarily continental, but includes portions of the Atlantic and Arctic Oceans. It is covers Europe and Asia.
4. **African Plate**: It is contains both continental and oceanic crust. It is includes African continent and surrounding oceanic crust.
5. **Antarctic Plate**: It is mostly continental, covering the Antarctic continent and surrounding oceanic crust.

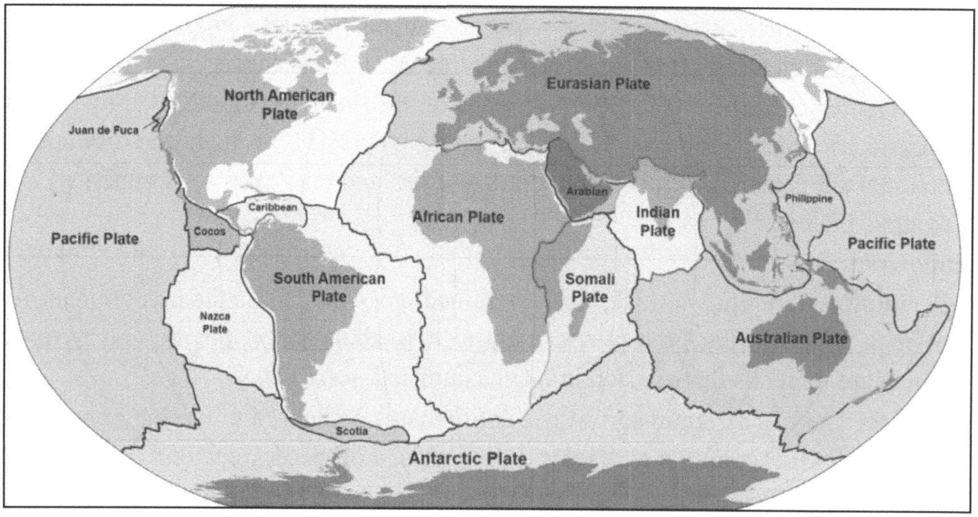

Figure 21.1: World Map of Tectonic Plates: Showing major and minor plates and their boundaries. (Reference: Wikipedia, The Free Encyclopedia.)

6. **Indo-Australian Plate**: It is a mix of the Indian and Australian plates, and it has crust that is both continental and oceanic in characteristics. It is long enough to reach all the way to Australia.
7. **South American Plate**: It includes both oceanic and continental crust in its composition. Specifically, it encompasses the continent of South America as well as portions of the Atlantic Ocean.

Minor Tectonic Plates

1. **Nazca Plate**: The oceanic plate in question is situated in the eastern part of the Pacific Ocean. It brings for the formation of the Andes Mountains through its interaction with the South American Plate.

2. **Philippine Sea Plate**: It is an oceanic plate that can be found in the western part of the Pacific Ocean. There are a number of subduction zones and active tectonic boundaries that extend throughout its perimeter.
3. **Cocos Plate**: There is an oceanic plate that can be found off the shore of Central America's west coast. It has interactions with the plates that are located in the Caribbean and North America.
4. **Caribbean Plate**: Both oceanic and continental crusts are included in its composition. The North American Plate and the South American Plate are both positioned in close proximity to it.
5. **Scotia Plate**: It is a rather small plate that can be found in the middle of the Atlantic and South American plates. Both oceanic and continental crusts are included in its composition.

Causes of Plate Movement

In the mantle and lithosphere of the Earth, there are a number of processes that are interrelated and are responsible for driving the movement of tectonic plates. These processes are responsible for the generation of the forces that are required to move the rigid plates over the surface of the Earth.

A. **Mantle Convection:** The major driving factor behind plate tectonics is mantle convection, as stated in the last sentence. The heat that emanates from the Earth's core causes the mantle, which is made up of rock that is of a semi-fluid consistency, to convect. The steps that are involved in this procedure are as follows:

1. Heat Source: Radioactive decay and residual heat from the formation of the globe both contribute to the generation of heat produced by the core of the Earth.
2. Convection Currents: The material that makes up the mantle becomes unstable as a result of heat, and it rises towards the lithosphere. It creates a cycle by becoming denser as it cools, which causes it to sink back towards the core of the object.
3. Plate Movement: The movement of the tectonic plates is caused by the convective flow in the mantle, which causes the base of the plates to be subjected to drag.

B. **Slab Pull:** The subduction zone is the location where a cold and dense oceanic plate sinks into the mantle, which results in the formation of a force known as slab pull.

1. Gravitational Pull: The sinking slab exerts a pull on the rest of the plate, drawing it towards the subduction zone.
2. Driving Force: Slab pull is considered one of the most significant forces driving plate motion.

C. **Ridge Push:** Ridge push, also known as gravitational sliding, occurs at mid-ocean ridges where new oceanic crust is formed. The process involves:

1. Formation of New Crust: At mid-ocean ridges, magma rises, which results in the formation of new oceanic crust.

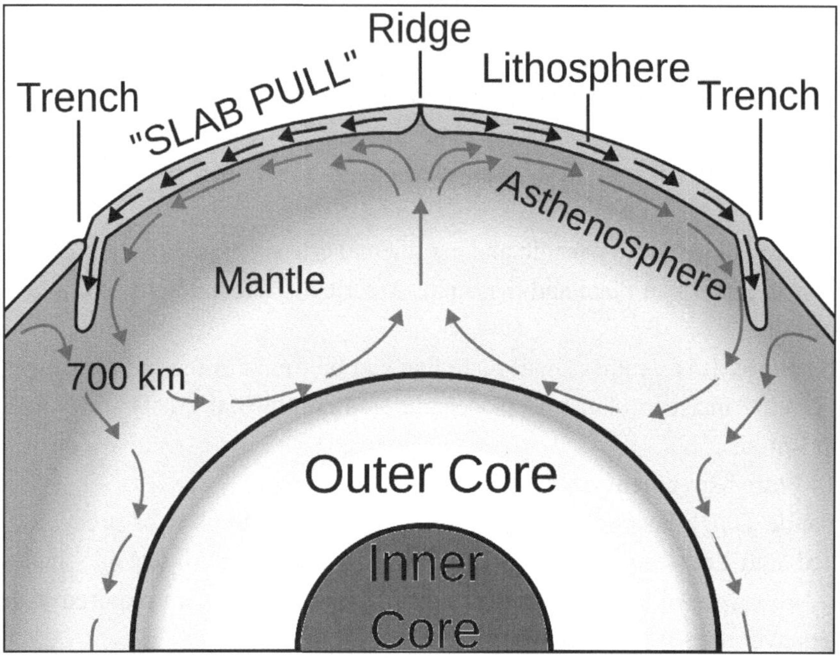

Figure 21.2: Diagram of Mantle Convection: Illustrating the process of heat-driven convection currents in the mantle. (Reference: *Wikipedia, the Free Encyclopedia.*)

2. Elevation of Ridge: The newly produced crust is heated and floating, which is what causes the ridge to be elevated on the surface.
3. Gravitational Force: As the crust cools and becomes denser, it slides down the slope of the ridge, which causes the plate to move away from the axis of the ridge.

D. Mantle Plumes and Hotspots: An upwelling of heated mantle material that results in the formation of volcanic hotspots is known as a mantle plume. During this process, you will:

1. Upwelling: A plume is formed when hot material from the mantle makes its way up from the depths of the earth.
2. Volcanic Activity: The plume travels to the lithosphere, where it causes a hotspot and is responsible for seismic activity.
3. Plate Motion Indicator: Plates move over hotspots, which provide a record of plate movement. Hotspots remain immobile while plates move over them.

Plate Interactions and Features

At their boundaries, the lithospheric plates of the Earth interact with one another, and the nature of this interaction determines the types of geological processes and landforms that can arise. The following is a list of the primary types of plate interactions and the characteristics that they have:

1. **Divergent Boundaries (Constructive Boundaries)**: The plates move at a distance from one another. Typically, this takes place in the crust of the ocean, but it can also take place in the crust of the continents.

Key Features:

A. **Mid-Ocean Ridges**: During the process of plate tectonics, magma from the mantle rises to fill the void, resulting in the formation of new oceanic crust. This results in the construction of mountain ranges that are located beneath the water and are referred to as mid-ocean ridges (for example, the Mid-Atlantic Ridge).

B. **Rift Valleys**: In the case of continental plates, divergence can result in the formation of rift valleys, which are regions where the land expands, becomes thinner, and ultimately separates. The East African Rift Valley is a good illustration of this.

C. **Volcanism**: Volcanic activity is caused by magma rising to the surface at divergent boundaries, which results in the formation of new crust material.

D. **Seafloor Spreading**: This process involves the formation of new oceanic crust at mid-ocean ridges, which then gradually drifts away, causing plate motion to occur.

E. **Example**: The Mid-Atlantic Ridge, which is spreading the Atlantic Ocean basin apart.

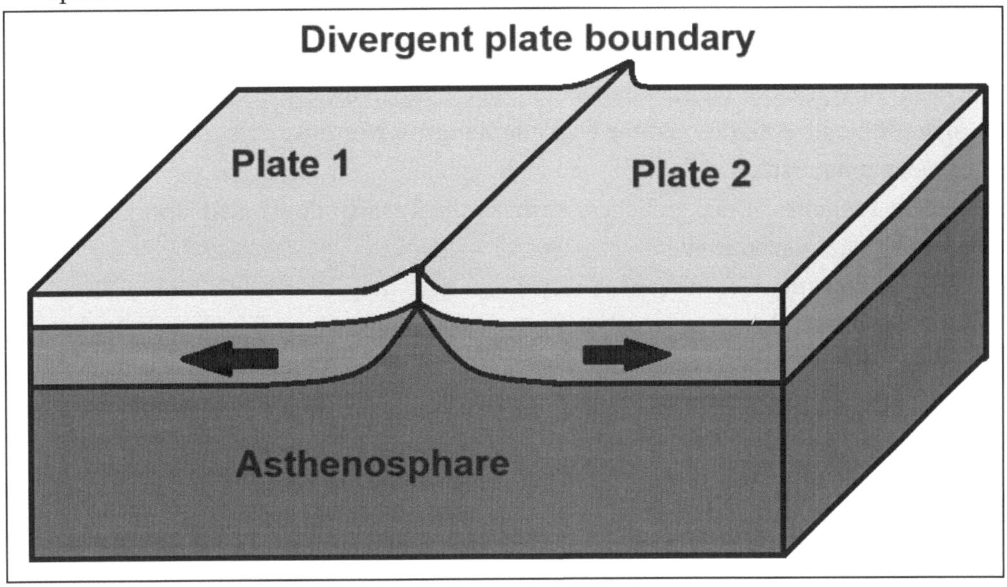

Figure 21.3: Diagrams of Divergent Plate Boundaries

2. **Convergent Boundaries (Destructive Boundaries)**: Plates go in the direction of one another. Both oceanic and continental plates, and a combination of the two, are capable of causing this phenomenon. There are three types of convergent boundaries:

A. **Oceanic-Continental Convergence:** In a process known as subduction, the oceanic plate, which is denser, is pushed beneath the continental plate, which is typically lighter. This ultimately results in the creation of volcanic arcs and deep ocean trenches on the plate that is overriding the ocean. Andes Mountains and the Peru-Chile Trench are two examples of such mountains.

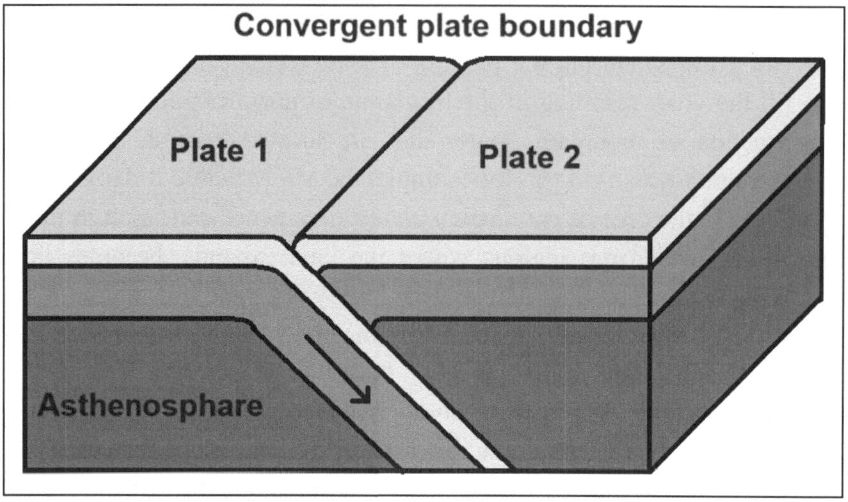

Figure 21.4: Diagrams of Convergent Plate Boundaries

Key Features:

a. **Volcanic Arc**: The melting of the subducting oceanic plate results in the formation of a chain of volcanoes on the continental plate with a volcanic chain. For instance, the Andes in South America and the Cascade Range in North America are both examples of mountain ranges.

b. **Ocean Trenches**: Deep trenches, such as the Peru-Chile Trench, form where the oceanic plate is subducted.

B. **Oceanic-Oceanic Convergence:** When two oceanic plates collide, one of them will subduct beneath the other, which will result in the formation of deep-sea trenches and volcanic island arcs. For instance, the Mariana Trench and the Mariana Islands are both examples.

Key Features:

- **Island Arc**: There is a series of volcanic islands that are bent and create a chain that runs parallel to an ocean trench. Islands such as the Mariana Islands and the Aleutian Islands are two specific examples.
- **Ocean Trench**: In the ocean floor, there are deep depressions, such as the Mariana Trench, which is the deepest part of the oceans on Earth.

C. **Continental-Continental Convergence:** Because of their equal densities, neither continental plate subducts when two collide. Rather, mountain ranges are formed as the crust folds and crumples. As an illustration, consider the Himalayas, which were created when the Indian and Eurasian plates collided.

Key Features:

Mountain Ranges:

The construction of massive mountain ranges, such as the Himalayas, is a result of the collision of two continental plates. The Indian Plate is colliding with the Eurasian Plate, which results in the formation of an enormous mountain range.

1. **Transform Boundaries (Conservative Boundaries)**: Along fault lines, plates slide past one another horizontally. At transform borders, the lithosphere is neither created nor destroyed.

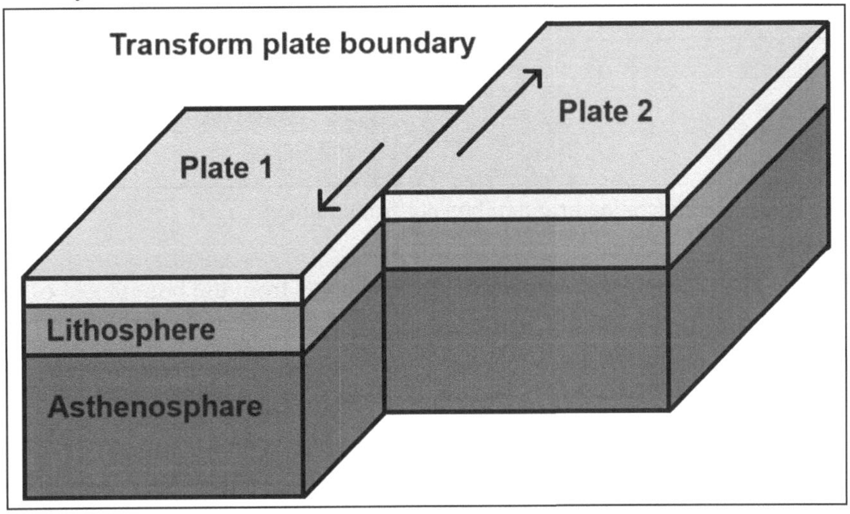

Figure 21.5: Diagrams of Transform Plate Boundaries

Key Features:

A. **Fault Lines**: The Earth's crust has large cracks or fractures where the plates slip past one another. The San Andreas Fault in California is the most well-known transform fault.

B. **Earthquakes**: Because the sliding plates might become trapped and stress accumulates until it is released as seismic activity, transform boundaries are frequently linked to earthquakes.

- **Example**: The San Andreas Fault, where the Pacific Plate and the North American Plate move past each other.

Note: J. Tuzo Wilson (1960s): The concept of transform faults—plate boundaries where two plates slide past one another—was first proposed by Wilson. He also contributed to the development of the theory that the lithosphere of Earth is composed of a number of inflexible plates known as tectonic plates, which move and interact at their borders. His contributions improved the understanding of plate tectonics.

Intraplate Activity: Some geological activity takes place within a plate, even though the majority of plate interactions take place at boundaries.

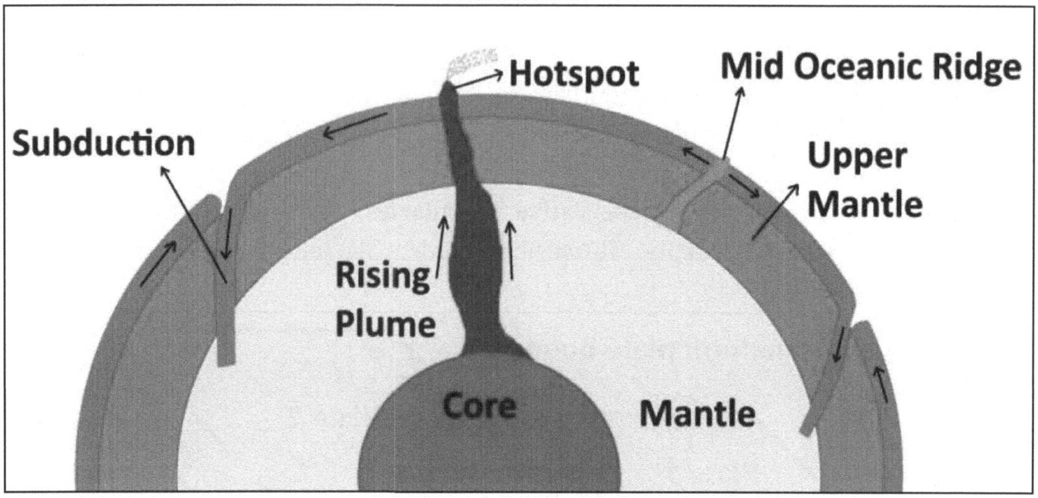

Figure 21.6: Diagrams illustration origin of hotspot from plume.

Key Features:

- **Hotspots**: The volcanic zones that are situated apart from the boundaries of the plates are created by plumes of hot mantle that are rising from the depths of the Earth. Isolated volcanic activity is produced by these plumes, which results in the formation of volcanic islands such as Hawaii. As the plate advances over the hotspot, a series of seamounts or volcanoes begins to form in successive layers.
- **Mantle Plumes**: In the process of rising rock from the deep mantle, which is characterized by extremely high temperatures, substantial volcanic activity can occur.

Rates of Plate Movement

The individual tectonic plate and the forces that are acting on it both have a significant impact on the rate of plate movement, which can vary greatly. In general, the velocity of plates can range anywhere from a few millimeters to several centimeters of movement every year.

A. Fast-Moving Plates

a. Pacific Plate: Travels 7–10 cm annually on average. Driven by ridge push atmid-ocean ridges and slab pull at subduction zones.

b. Nazca Plate: Moves at a rate of about 7 cm/year. Subducting beneath the South American Plate, contributing to the formation of the Andes.

B. Slow-Moving Plates

a. Eurasian Plate: Moves between one and two centimeters every year on average. Mostly influenced by interactions with nearby plates.

b. African Plate: Travels at a speed of roughly 2.5 cm annually. Distinguished by interactions with the neighboring plates and different borders.

Factors Influencing Plate Velocity

Tectonic plate movement is influenced by a number of variables. These elements include the underlying mantle's composition, the plate borders' interactions, and the thickness and size of the plates.

C. Plate Size and Thickness

a. Large Plates: Because of their increased mass and the pull that the mantle exerts, larger plates often move more slowly.

b. Thin Plates: Plates that are thinner may move more quickly because they have less resistance to the forces that are causing their motion.

D. Nature of the Mantle

a. Viscosity: The viscosity of the mantle has an effect on the rate of convection, which in turn has an effect on the pace at which plates move. Both convection and plate motion can occur more quickly when the mantle has a lower viscosity.

b. Temperature: Higher temperatures in the mantle cause a decrease in viscosity and an increase in convective flow, which speeds up horizontal plate movement.

Measuring Plate Movement

There are a few different approaches that scientists take in order to quantify and calculate the rate of plate movement. Methodologies such as geological and geophysical approaches, in addition to contemporary satellite-based measurements, are included in these methodologies.

A. Geological and Geophysical Techniques

a. Paleomagnetism: The investigation of the magnetic field of the Earth as it is recorded in rocks. A record of plate movement throughout time can be found on the ocean floor in the form of magnetic stripes, which are generated by the spreading of the seabed.

b. Radiometric Dating: The process of dating volcanic rocks and sediments enables scientists to determine the age of geological features and to estimate the rate at which plates are moving.

c. Seafloor Spreading Rates: Scientists are able to determine the rate of seafloor spreading and plate movement by measuring the distance between magnetic stripes on the ocean floor that are known to be of a certain age.

B. Satellite-Based Measurements

a. Global Positioning System (GPS): Real-time, accurate measurements of the location of points on the surface of the Earth are made possible by the Global Positioning System (GPS). It is possible for scientists to determine the rate of plate movement as well as the direction of movement by monitoring these spots throughout time.

b. Satellite Laser Ranging (SLR): Satellite-based remote sensing (SLR) calls for the use of laser pulses to determine the distance between satellites and fixed sites on the surface of the Earth. Changes in the shape of the Earth and movements of its plates can be monitored with the use of this technology.

c. Synthetic Aperture Radar (InSAR): InSAR is able to detect ground displacement and deformation by utilizing radar images from satellites. This provides essential information regarding plate movements and tectonic activity.

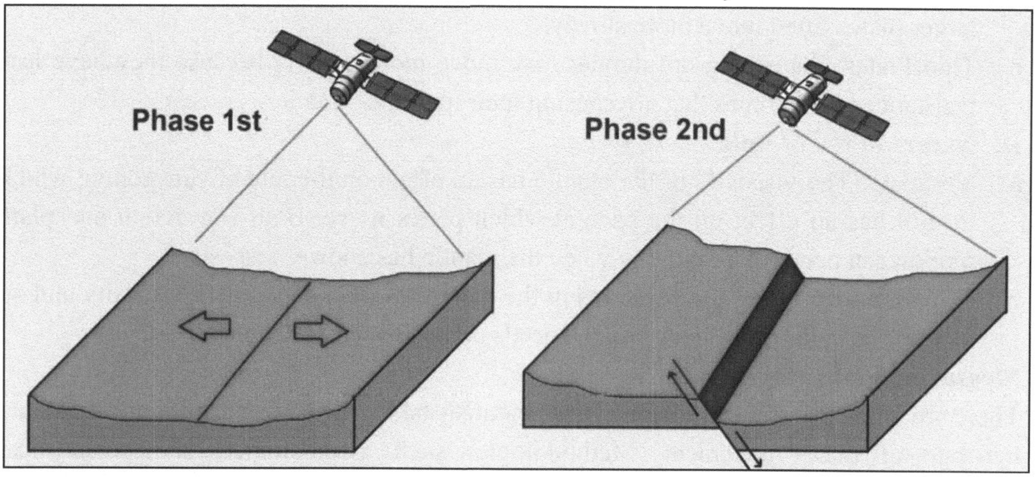

Figure 21.7: Satellite Measurement Techniques: Cartoon depiction of the InSAR methodology. Two SAR images of the same area are acquired at different times.

Implications of Plate Movement Rates

1. **Earthquake Frequency and Magnitude**

A. **Faster plate movement:**Plate movement that occurs at a faster rate can lead to earthquakes that occur more frequently and with greater intensity along tectonic borders, particularly at transform and convergent boundaries. This phenomenon occurs near transform boundaries, such as the San Andreas Fault, where plates that are moving at a quicker rate allow stress to accumulate at a faster rate, resulting in frequent and significant earthquakes. At subduction zones, such as the Pacific and Nazca Plates, the release of energy occurs when the plates ultimately slip, resulting in larger and more violent earthquakes. This occurs because quicker subduction causes the plates to slip more quickly.

B. **Slower plate movement:** Plate movement that is slower can still cause big earthquakes, although they may occur less frequently due to the gradual accumulation of stress rather than the sudden occurrence of earthquakes.

2. **Volcanic Activity**

A. **Convergent boundaries**: The rate at which magma is formed and brought to the surface is influenced by the speed of plate movement in subduction zones, which are regions where oceanic plates sink beneath continental plates. The Pacific Ring of Fire is one of the locations that potentially experience an increase in the frequency of volcanic eruptions as a result of faster global convergence.

B. **Divergent boundaries**: The rate of seafloor spreading is increased when there is a faster divergence of plates (for example, at mid-ocean ridges), which also encourages continuous volcanic activity as new crust is created.

3. **Mountain Building and Uplift Rates:**The rate at which mountains are formed is directly influenced by the speed at which plates are starting to converge. During the occurrence of continental-continental convergent borders, such as the collision between the Indian Plate and the Eurasian Plate, the occurrence of quicker plate movement results in the production of mountains and the uplift of mountains faster. The Himalayas serve as a prime illustration of the convergence of the Indian Plate and the Eurasian Plate, which is occurring at a pace of approximately 5 centimeters per year. This convergence is causing the mountain range to continue to rise.In places that move at a slower pace, the processes of uplift and mountain-building are more gradual, and changes in topography may take place over the course of millions of thousands of years.
4. **Formation of Ocean Basins and Seafloor Spreading**: At divergent boundaries, such as the Mid-Atlantic Ridge, the rate of seafloor spreading is determined by the movement rates of the plates. Divergence rates that are faster are associated with a greater rate of expansion of ocean basins, whereas divergence rates that are slower can restrict the formation of new oceanic crust.Throughout the course of geological time, the relative velocity of plate movement has the potential to influence the dimensions of ocean basins and the configuration of continents. As the North American and Eurasian plates continue to separate, the Atlantic Ocean, for instance, continues to spread out and become more expansive.
5. **Subduction and Recycling of Oceanic Crust**: The process of subduction into the mantle occurs more quickly for plates that move more quickly, particularly oceanic plates. Because of the greater subduction rates, the lifespan of oceanic crust is often shorter than that of continental crust. This process has an effect on the longevity of oceanic crust.A faster subduction process can result in a more rapid recycling of crust, which in turn can lead to an increase in the occurrence of volcanic arcs and the construction of deep-sea trenches.
6. **Impacts on Global Climate**: Plate tectonics has the ability to affect long-term climate changes by contributing to the modification of ocean currents, mountain ranges, and patterns of air circulation. Plate movements that occur at a faster rate can result in more significant adjustments in these parameters over the course of time.The disintegration of supercontinents, such as Pangaea, and the movement of plates have the potential to influence the carbon cycle on a worldwide scale. Rapid plate motion

has the ability to enhance volcanic activity, leading to the emission of a greater quantity of carbon dioxide into the atmosphere. This phenomenon has the potential to influence world temperatures over geological timescales.

7. **Hazard Assessment and Prediction**: In order to forecast regions that are prone to earthquakes, volcanic activity, and tsunamis, scientists have the ability to gain a better understanding of the movement speeds of tectonic plates. For example, regions where plates move quickly, such as the Pacific Ring of Fire, are more susceptible to frequent and intense seismic and volcanic events. The monitoring of the speed of plate motion, especially near fault lines and subduction zones, is of utmost importance in order to forecast the occurrence of future natural disasters and to provide early warning systems for populations that are at vulnerability.
8. **Geological Evolution and Continental Drift**: Over the course of millions of years, the relative movement rates of tectonic plates are the driving force behind the process of continental drift, which involves the slow shifting of the position of continents. As a result of this movement, the arrangement of continents, the creation of supercontinents (such as Pangaea), and the opening and shutting of ocean basins are all influenced.When it comes to the process of continental rearrangement, plates that move more quickly speed up the process, whereas plates that move more slowly delay substantial changes in the shape of the Earth's surface area.

Case Studies of Plate Movement Rates

1. **Pacific Plate**: At a rate of approximately 7–10 centimeters per year, the Pacific Plate is one of the plates that move the quickest. There has been a significant amount of volcanic activity and earthquakes along the Pacific Ring of Fire as a result of this rapid travel. Intense volcanic activity, such as that which occurs in the Japan Trench, is caused by the subduction of the Pacific Plate beneath the plates that surround it. This subduction also results in large earthquakes, such as the Tōhoku earthquake that occurred in 2011.
2. **Indian Plate**: The Indian Plate is heading in the direction of the Eurasian Plate at a rate of around 5 centimeters each year. The increasing elevation of the Himalayas and the occurrence of regular earthquakes, such as the one that occurred in Nepal in 2015, are both consequences of global warming. The landscape of the region has been drastically reshaped as a result of the rapid convergence.
3. **Mid-Atlantic Ridge**: Along the Mid-Atlantic Ridge, the difference between the Eurasian and North American plates is increasing at a pace of around two to four centimeters per year. A relatively constant and moderate expansion of the Atlantic Ocean is produced as a result of this slower rate of movement within the ridge, which also results in continuing volcanic activity.

Evidence Supporting Plate Tectonics

A. **Seafloor Spreading:** Harry Hess was the one who made the discovery that seafloor spreading takes place along mid-ocean ridges, which are the locations where new oceanic crust is created and extends outward. There is evidence of seafloor spreading that can be found in the form of magnetic stripes on the ocean floor. These stripes record reversals in the magnetic field of the Earth.

B. **Distribution of Earthquakes and Volcanoes:** Plate boundaries are the locations where the majority of earthquakes and volcanoes occur, which lends support to the concept of plate interactions. The Pacific Plate is surrounded by the Ring of Fire, which is a region that is characterized by frequent earthquakes and volcanic activity.

C. **Matching Geological Features:** Rock formations, mountain ranges, and geological features that are similar to one another and may be found on other continents lend support to the concept of previous continental links. The Appalachian Mountains in North America and the Caledonian Mountains in Europe and Greenland are examples.

D. **Fossil Evidence:** Similar fossils discovered on continents that are currently separated by oceans provide evidence that these continents were previously related to one another. Examples include fossils of the Mesosaurus, found in both South America and Africa.

E. **Paleoclimatic Evidence:** The concept of migrating continents is supported by evidence. This evidence includes glacial deposits that have been found in locations that are currently tropical. Coal beds in Antarctica indicate it was once in a warmer climate near the equator.

The appearance of the concept of plate tectonics has fundamentally altered our understanding of the geology of the Earth. The movement of tectonic plates and the interactions between them are explained, which enables one to gain an understanding of the processes that led to the development of continents, ocean basins, mountain ranges, earthquakes, and volcanic activity. The field of plate tectonics is always developing, which enables researchers to gain fresh insights into the history, present, and future of the planet Earth.

Chapter 22
Volcanoes

Introduction to Volcanoes

Volcanoes are geological formations that are formed when magma from the Earth's mantle escapes to the surface. This results in eruptions that can vary in intensity and duration, and they are characterized by both of these characteristics. These eruptions result in a wide variety of landforms and have a profound impact on the environment from where they originate. The understanding of the many types of volcanoes, the landforms that they produce, and their global distribution is vital for grasping the role that volcanoes play in the geology of the Earth and reducing the risks that are associated with volcanic activity.

Types of Volcanoes

It is possible to categorize volcanoes according to their shape, the manner in which they erupt, and the kinds of materials that they release. Shield volcanoes, stratovolcanoes, cinder cone volcanoes, and lava domes are the categories that make up the majority of volcanoes.

A. Shield Volcanoes

- Shape: Sides that are broad and gently sloped, mimicking the shield used by a warrior.

- Eruption Style: In most cases, effusive eruptions are characterized by the production of basaltic lava with a low viscosity, which flows over extensive distances.
- Materials: Primarily basaltic lava with occasional pyroclastic material.
- Examples

a. Mauna Loa (Hawaii): One of the largest shield volcanoes on Earth, known for its frequent but non-explosive eruptions.

b. Kilauea (Hawaii): Continuously erupting since 1983, providing valuable insights into effusive volcanic activity.

B. Stratovolcanoes (Composite Volcanoes)

- Shape: Consisting of alternating layers of lava flows, volcanic ash, and other pyroclastic materials, these cones have steep sides and are symmetrical in shape by nature.
- Eruption Style: In the presence of andesitic to rhyolitic magma, which has a high viscosity and has the ability to trap gasses, explosive eruptions can occur.
- Materials: Additionally, pyroclastic flows and ash are present in addition to andesite, dacite, and rhyolite lavas.
- Examples

a. Mount St. Helens (USA): Known for its catastrophic eruption in 1980 that significantly altered its structure.

b. Mount Fuji (Japan): An iconic stratovolcano with a history of explosive eruptions.

C. Cinder Cone Volcanoes

- Shape: Small cones with steep sides, comprising primarily of volcanic ash, tephra, and cinders have been observed.
- Eruption Style: The eruptions that are relatively short-lived and create pyroclastic debris to be released.
- Materials: Basaltic to andesitic cinders and volcanic ash.
- Examples

a. Parícutin (Mexico): Emerged in a farmer's field in 1943 and grew rapidly over a few years.

b. Sunset Crater (USA): A well-preserved cinder cone in Arizona.

D. Lava Domes

- Shape: In the process of slowly extruding viscous lava, dome-shaped mounds come into existence with steep sides.
- Eruption Style: Effusive eruptions are characterized by the production of thick and sticky lava, which ultimately results in the formation of domes.
- Materials: Lava that ranges from rhyolitic to dacitic.
- Examples

a. Lassen Peak (USA): A large lava dome that last erupted in 1915.
b. Mount St. Helens Lava Dome (USA): Formed within the crater left by the 1980 eruption.

E. Other Types

1. Calderas

- Description: As a result of a big eruption, large depressions that resemble basins are generated as a volcano collapses.
- Examples: Yellowstone Caldera (USA), Crater Lake (USA).

2. Fissure Volcanoes

- Description: In the case of linear volcanic vents, lava is released through these vents, which typically lack a central peak.
- Examples:Laki (Iceland), Eldgjá (Iceland).

On the basis of degree of explosive activity and nature of eruption

A detailed continuity from low-intensity to extremely violent eruptions is provided by this classification. This progression is based on the degree of explosive activity and the composition of the material that is erupted to determine the classification. Based on the descriptions you provided, the following is a more structured breakdown:

1. Hawaiian Type

- Explosivity: Either nonexistent or minute.
- Nature of Eruption: A quiet and effusive eruption is characterized by the smooth flow of lava from cracks or vents into the surrounding environment.
- Lava Characteristics: The characteristics of lava include low viscosity basaltic lava, which enables the flow of lava over extensive extended distances.
- Eruption Frequency: There is a frequent yet gentle occurrence of the eruption.
- Example: Kīlauea in Hawaii.

2. Strombolian Type

- Explosivity: Mild to moderate.
- Nature of Eruption: Lava fountains and minor explosions are the direct result of periodic eruptions of gas-rich magma, which are characterized by their nature. In most cases, eruptions are brief and rhythmic in quality.
- Lava Characteristics: The characteristics of lava include that it is more in viscosity than Hawaiian eruptions, yet it is still quite fluid
- Eruption Frequency: The frequency of eruptions is characterized by periodic intervals of brief, explosive bursts.
- Example: Stromboli volcano, Italy.

3. Vulcanian Type

- Explosivity: Extremely moderate to intense.

- Nature of Eruption: When compared to Strombolian eruptions, the nature of the eruption is stronger and occurs less frequently. As a result of the increased viscosity of the lava, there is an increase in pressure. In the event of an explosion, ash, bombs, and volcanic gasses are released into the atmosphere.
- Lava Characteristics: The characteristics of lava include a high viscosity, which causes it to solidify rapidly, culminating in the formation of explosive ash clouds.
- Eruption Frequency: Eruption frequency refers to the length of time that passes between eruptions.
- Example: Sakurajima, Japan.

4. Vesuvian Type

- Explosivity: Extremely explosive qualities.
- Nature of Eruption: Large eruptions that occur after longer periods of dormancy (often measured in decades) are referred to as the nature of eruptions. Eruptions release enormous quantities of ash, gasses, and pyroclastic material into the atmosphere. Flows of lava might or might not take place.
- Lava Characteristics: In terms of characteristics, lava is characterized by its high viscosity and significant explosive force.
- Eruption Frequency: The frequency of eruptions is characterized by long periods between eruptions, which can be measured in tens of years.
- Example: Mount Vesuvius, Italy.

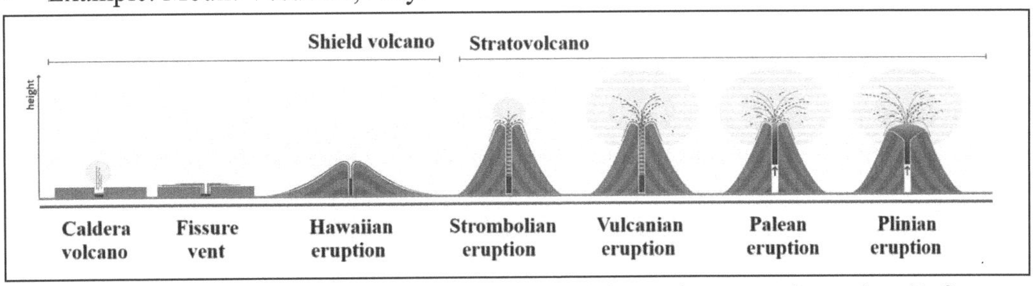

Figure 22.1: On the basis of degree of explosive activity and nature of eruption (Reference: Wikipedia, The Free Encyclopedia)

5. Plinian Type

- Explosivity: Highly aggressive and violent
- Nature of Eruption: The nature of the eruption is marked by long eruption columns that can reach the stratosphere, and it is a more extreme variation of the Vesuvian eruptions. Ashes pumice, and gasses are released in enormous quantities, which frequently results in the formation of pyroclastic flows.
- Lava Characteristics: Fragmental products predominate, and there is either very little or no flow of lava.

- Eruption Frequency: As a result of prolonged periods of dormancy, the frequency of eruptions is extremely rare.
- Example: Mount Vesuvius (AD 79), Mount St. Helens (1980).

6. **Pelean Type**

- Explosivity: The most violent behaviour.
- Nature of Eruption: One of the characteristics of an eruption is the formation of pyroclastic flows, also known as nuéesardentes, which are characterized by flaming clouds of ash and gas. It's possible for lava domes to collapse, which would set off pyroclastic flows that travel at high speeds and destroy everything in their path.
- Lava Characteristics: Lava is characterized by its viscous nature, giving rise to domes that have the potential to collapse and cause explosive explosions.
- Eruption Frequency: It is a rare occurrence, but it causes a significant amount of destruction.
- Example: Mount Pelée (1902), Soufrière Hills.

With rising volcanic intensity and various types of ejecta (lava, ash, gasses, and pyroclastic debris), this classification system illustrates a development from non-explosive, effusive eruptions to highly destructive and explosive ones. This trend is highlighted by the fact that the strength of the eruptions increases.

Volcanic Landforms

The activity of volcanoes results in the production of a wide variety of landforms, each of which has its own unique characteristics and formation processes.

A. Lava Plateaus

1. **Description**

- Formation: In the process of formation, vast lava flows are responsible for covering large areas and subsequently solidifying into broad, flat plains.
- Materials: Predominantly composed of basaltic lava composition.

2. **Examples**

- Columbia River Plateau (USA): Formed by repeated basaltic lava flows.
- Deccan Traps (India): One of the largest volcanic features on Earth, created by massive basaltic eruptions.

B. Volcanic Cones

1. **Shield Volcanoes**

- Characteristics: Lava flows with low viscosities that have developed sides that are broad and gently sloped.
- Examples: Mauna Loa (Hawaii), Kilauea (Hawaii).

2. **Stratovolcanoes**

- Characteristics: Cones with steep sides and symmetrical shapes, created by alternating layers of pyroclastic material and lava at different levels.
- Examples: Mount Fuji (Japan), Mount Rainier (USA).

3. Cinder Cones

- Characteristics:The cones are small and steep-sided, and they are primarily formed of pyroclastic debris.
- Examples:Parícutin (Mexico), Sunset Crater (USA).

C. Calderas

- Formation: The collapse of a volcano after a massive eruption results in the formation of large depressions under the surface.
- Characteristics: Caldera lakes are formed when calderas are filled with water.
- Examples

a. Yellowstone Caldera (USA): A supervolcano with a history of massive eruptions.
b. Crater Lake (USA): Formed by the collapse of Mount Mazama.

D. Lava Domes

- Formation: Constructed through the gradual extrusion of thick lava, which leads to the formation of domes with steep sides.
- Characteristics: In the event that the dome collapses, it has the potential to generate pyroclastic flows.
- Examples

a. Lassen Peak (USA): A large lava dome.
b. Mount St. Helens Lava Dome (USA): Formed within the crater left by the 1980 eruption.

E. Volcanic Plains

- Formation: Extensive regions that are typically basaltic and are covered by lava flows.
- Characteristics: Plains that is open and relatively flat.
- Examples

a. Snake River Plain (USA): Formed by repeated basaltic lava flows.
b. Icelandic Volcanic Plains: Created by extensive volcanic activity.

Distribution of Volcanoes

Worldwide, volcanoes are dispersed in an unequal manner, with the majority of them being located near the boundaries of tectonic plates.

A. Plate Tectonics and Volcanic Distribution

1. Divergent Boundaries

- Description: Plates undergo a geographical separation, which enables magma to ascend and form a new crust.

- Examples: Mid-Atlantic Ridge, East African Rift.

2. **Convergent Boundaries**

- Description: As the plates move closer to one another, one of them is pushed beneath the other, which results in the creation of magma and melting of the plates.
- Examples: Pacific Ring of Fire, Andean Volcanic Belt.

3. **Transform Boundaries**

- Description: Plates are able to slide past one other; however volcanism is not as prevalent in this region.
- Examples: San Andreas Fault (USA) has limited volcanic activity.

4. **Hotspots**

- Description: Volcanic activity that takes place away from plate boundaries and is generated by plumes of hot mantle material that emerge from the earth's interior.
- Examples: Hawaiian Islands, Yellowstone (USA).

B. Major Volcanic Belts and Regions

1. **Pacific Ring of Fire**

- Description: A significant region located within the basin of the Pacific Ocean, characterized by a significant number of volcanoes and a high level of seismic activity.
- Countries Affected: The west coastlines of North and South America, as well as Indonesia, Japan, the Philippines, and New Zealand are included in this category.

2. **Mid-Atlantic Ridge**

- Description: The Atlantic Ocean is characterized by a diverging boundary that runs down the center of the ocean.
- Countries Affected: When it comes to volcanic activity along the ridge, Iceland stands out as a remarkable example.

3. **Mediterranean Belt**

- Description: The regions surrounding the Mediterranean Sea that are filled with volcanic activity are included.
- Countries Affected: Italy (Mount Etna and Vesuvius), Greece (Santorini), and other locations.

4. **East African Rift**

- Description: There is a diverging border that runs through the African continent.
- Countries Affected: Mt. Kenya, Ethiopia (Erta Ale), and Ethiopia (Erta Ale).

5. **Hawaiian-Emperor Seamount Chain**

- Description: The Pacific Plate, which is moving over a hotspot, is responsible for the formation of a sequence of volcanic islands and seamounts.
- Notable Volcanoes**:** Kilauea, often known as Mauna Loa, is located in Hawaii.

Volcanic Hazards and Mitigation

A. Volcanic Hazards

1. Lava Flows

- Description: Rock that is already molten and flows out of a volcano.
- Impact: It has the ability to destroy everything that is in its path, but it normally moves at a pace that allows people to take shelter.

2. Pyroclastic Flows

- Description: Moving currents of hot gas and volcanic matter that move at a rapid pace.
- Impact: Extremely hazardous and capable of destroying anything that it encounters in their route.

3. Ashfall

- Description: The ash that is released from a volcano during an eruption and then falls back down to the ground.
- Impact: It is possible for it to cause respiratory problems, taint water supplies, and cause harm to machinery.

4. Lahars

- Description: As a result of the melting of snow and ice that occurs during an eruption, volcanic mudflows or debris flows are produced.
- Impact: In their route, they have the ability to bury communities and infrastructure.

5. Volcanic Gases

- Description: When an eruption occurs, gases such as sulfur dioxide, carbon dioxide, and hydrogen sulfide are discharged into the atmosphere.
- Impact: Possibly contributes to climate change and causes difficulties with the respiratory system.

B. Mitigation Strategies

1. Monitoring and Prediction

- Seismic Monitoring: Eruptions frequently start by earthquakes, which can be detected.
- Gas Emissions: It is possible to detect the presence of rising magma by monitoring the discharge of volcanic gases.
- Ground Deformation: Changes in the morphology of a volcano can serve as an indication of the movement of magma.

2. Evacuation Plans

- Preparedness: Planning and rehearsing evacuation strategies for communities located in close proximity to volcanoes.
- Education: Informing the general population of the potential dangers posed by volcanoes and the procedures to be followed in case of an emergency.

3. Building Codes

- Design: Buildings that are constructed to withstand the effects of volcanic hazards, such as ashfall.
- Location: A construction project should be avoided in high-risk locations such as lahar paths.

4. Hazard Mapping

- Zoning: In order to identify locations that are susceptible to various volcanic risks, it is necessary to create hazard maps.
- Land Use Planning: Implementing hazard maps as a means to direct development and mitigate risk.

Volcanoes are natural features that are both active and strong, and they have the ability to modify the surface of the Earth and influence the environment. To better manage volcanic hazards and reduce the negative effects of volcanic eruptions, it is helpful to have a better understanding of the many types of volcanoes, the landforms they produce, and their global distribution. As a result of technological advancements in monitoring and prediction, our capacity to forecast volcanic eruptions and safeguard communities that are at risk has significantly improved. Through the study of volcanoes, we are able to get an understanding of the inner workings of the Earth as well as the processes that have created our planet over the course of geological time.

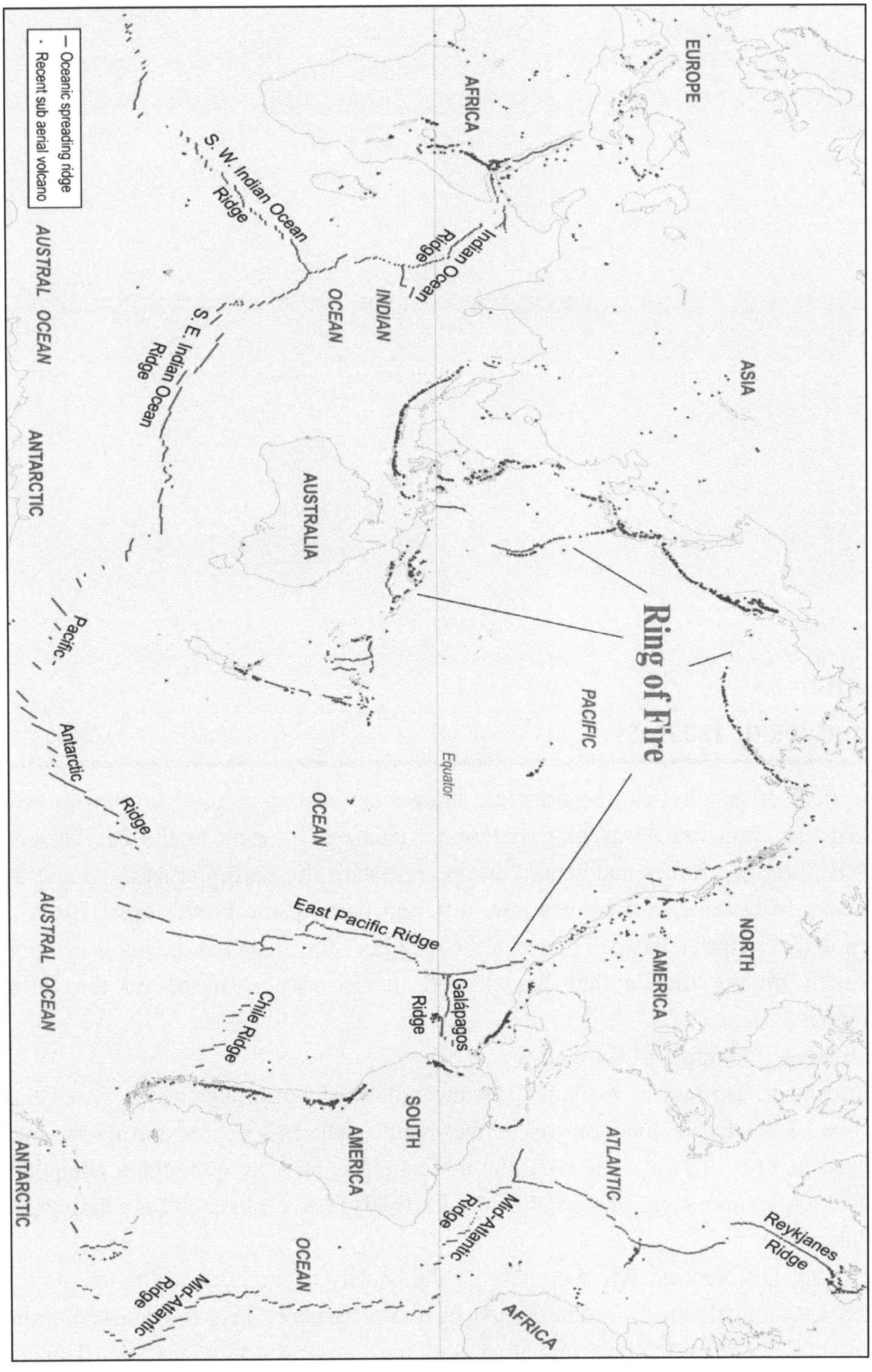

Oceanic spreading ridge
Recent sub aerial volcano
EUROPE
AFRICA
ASIA
S. W. Indian Ocean Ridge
Indian Ocean Ridge
INDIAN OCEAN
S. E. Indian Ocean Ridge
AUSTRALIA
Ring of Fire
PACIFIC OCEAN
Equator
Pacific Antarctic Ridge
East Pacific Ridge
Galápagos Ridge
Chile Ridge
NORTH AMERICA
SOUTH AMERICA
ATLANTIC OCEAN
Mid-Atlantic Ridge
Mid-Atlantic Ridge
Reykjanes Ridge
AFRICA
AUSTRAL OCEAN
ANTARCTIC
AUSTRAL OCEAN
ANTARCTIC

Chapter 23
Principles of Isostasy

In the field of geophysics, the principle known as Isostasy is used to explain how the crust of the Earth maintains its gravitational balance. It refers to the equilibrium that exists between the forces that cause certain regions of the Earth's crust to be elevated or depressed. In essence, it describes how different parts of the Earth's crust "float" at an elevation that strikes a balance between the weight of the crust and the buoyant force that is exerted by the mantle that lies beneath it. Isostasy is based on the following fundamental ideas.

Archimedes' Principle of Buoyancy

- Buoyancy: Isostasy is explained by the notion of buoyancy, which states that the crust of the Earth floats on top of the mantle, which is denser and more fluid-like than the crust. In the same way that an iceberg is buoyant although a component of its bulk is submerged, the crust of the Earth also has a piece that is "submerged" in the mantle.
- Density Differences: When comparing the density of the crust to that of the mantle, the level of "floating" is determined by the difference. This is because continental crust is less dense than oceanic crust, which is denser. Continental crust floats higher than oceanic crust.

Gravitational Equilibrium

- Hydrostatic Equilibrium: An instance of the crust being in Isostatic balance occurs when the forces that are responsible for elevating certain areas of the crust, known as buoyant forces, are equal to the forces that cause those parts to sink, which include gravity and the weight of the material that is overlaying them. Vertical movements, also known as Isostatic adjustments, are the means by which regions that are out of equilibrium adapt over time.
- Restoring Forces: As a result of a disturbance, like as erosion or glacial, the crust undergoes movement in either direction, either upward or downward, in order to regain equilibrium.

Types of Isostasy Models

There are some models that describe Isostatic balance based on crustal thickness and composition:

1. **Airy Isostasy**

- Assumption: There is a consistent density across the Earth's crust, despite its variable thickness.
- Mechanism: Mountains have deeper "roots" that allow them to extend further into the mantle, which is the reason for their higher elevation. In a similar manner, low-lying regions have a crustal root that is thinner.

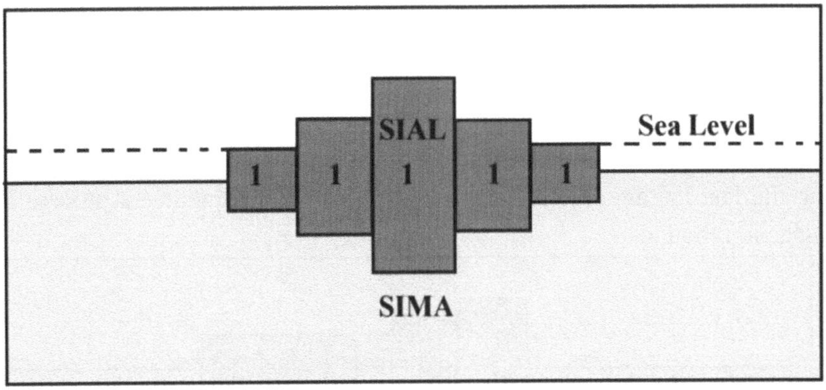

Figure 23.1: Airy Isostasy

- Example: The Himalayas have deep roots, which support their large surface mass. According to Airy's model, the thick crust beneath the mountains compensates for their high elevation.

2. **Pratt Isostasy**

- Assumption: Although the thickness of the Earth's crust is consistent, the density of the crust changes.

- Mechanism: Those portions of the crust that are less dense tend to sink and produce lower altitudes, while those that are more dense tend to rise and form higher elevations.
- Example: Mid-ocean ridge, where the less dense crust rises higher than the surrounding denser oceanic crust, is explained by Pratt Isostasy.

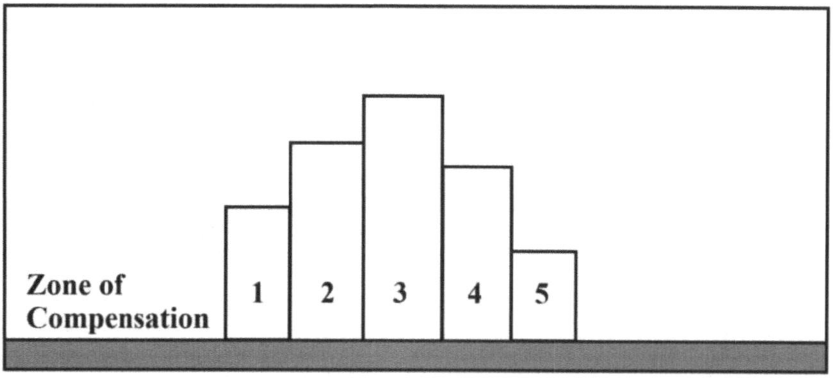

Figure 23.2: Pratt Isostasy

3. **VeningMeinesz or Flexural Isostasy**

- Assumption: Rather of being composed of hard blocks, the lithosphere of the Earth has the characteristics of a plate that is flexible and elastic.
- Mechanism: Because to the presence of loads on the crust, such as mountains or ice sheets, the lithosphere undergoes bending and flexing over a substantial area. The process by which the lithosphere is able to withstand huge loads without instantly sinking is explained by this concept.
- Example: Greenland's lithosphere, which was depressed by the weight of ice sheets during the last ice age, is now rising slowly as the ice melts—a process known as post-glacial rebound.

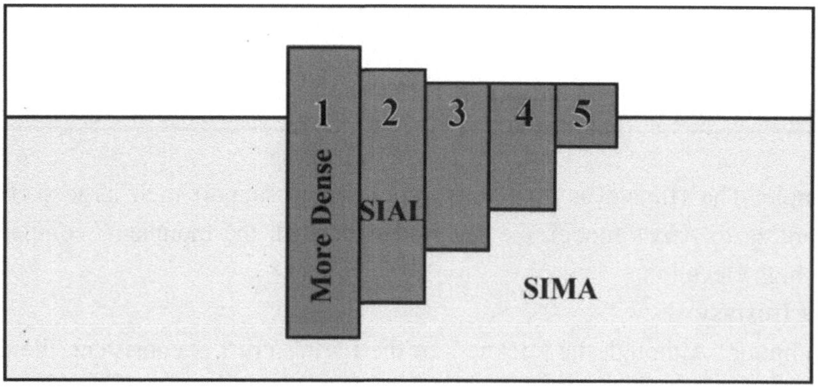

Figure 23.3: Heiskanen'sIsostasy

4. Heiskanen's Isostasy

- Assumption**:** The Isostatic balance of the Earth's crust is maintained by a mixture of factors, including fluctuations in density within the crust itself and the presence of deep roots beneath elevated landforms, such as mountains.
- Mechanism**:** Heiskanen's model provides an explanation for the fact that huge topographic features, like mountains, are sustained by both deeper roots in the crust (as is the case with Airy's model) and fluctuations in density (as is the case with Pratt's model). These components work together to establish equilibrium, with denser sections of the crust being able to compensate for less depth and lighter parts of the crust being compensated by deeper roots. A combination of these elements is what brings about the equilibrium.
- Example**:** The Andes mountains have deep crustal roots and regional density variations, both contributing to their isostatic balance. The root depth under the mountains and differences in crustal density help support their massive elevation while maintaining equilibrium.

Post-Glacial Rebound (Isostatic Adjustment)

- Definition: When a significant weight, such as a glacier, is removed from a region, the crust that was previously depressed begins to rise and rebalance itself. This phenomenon is referred to as Isostatic rebound.
- Example: Scandinavia and Canada, where glaciers once weighed down the crust, are still experiencing uplift today after the ice melted thousands of years ago.

Erosion and Sedimentation

- Erosion: The process of material erosion from a mountain range results in a reduction in weight, which in turn enables the crust beneath to rise and maintain its Isostatic equilibrium.
- Sedimentation: On the other hand, as sediments collect in a basin or offshore, the additional weight forces the crust to sink in order to maintain equilibrium.
- Dynamic Adjustment: These processes serve as evidence of the continuous adjustments that the crust of the Earth undergoes in response to the redistribution of mass.

Oceanic vs. Continental Isostasy

- Oceanic Crust: Because it is denser and thinner than the mantle, the oceanic crust "floats" lower on the mantle, which is how it produces the ocean basins.
- Continental Crust: Because the continental crust is less solid and thicker, it will "float" higher, resulting in the formation of landmasses. This provides an explanation for the disparity in elevation that exists between ocean floors and continents.

Isostatic Compensation

- Isostatic Compensation: When a load is introduced, such as a glacier or a mountain, the material that is at the bottom of the mantle moves out from beneath it. Conversely, when the stress is removed, the material flows back. Maintaining equilibrium throughout time is made easier by this mechanism.
- Flow of Mantle Material: The mantle, also known as the asthenosphere, flows slowly in reaction to surface stresses, which helps accomplish isostatic compensation. The lithosphere, on the other hand, can bend or flex with the force of gravity.

Role of Isostasy in Mountain Building

- Orogeny (Mountain Formation): On the occasion of a collision between two tectonic plates, mountain ranges are elevated. By following to the laws of Isostasy, the deeper roots that are produced beneath the mountain contribute to the equilibrium of the bulk of the mountain that is located above it.
- Long-Term Stability: Isostasy is responsible for the long-term stability of the Earth's surface throughout geological time scales. It does this by preserving equilibrium in spite of the influence of tectonic and erosional processes.

Isostatic Disequilibrium

- Imbalance: It is observed that when the crust of the Earth is not in a state of Isostatic equilibrium, stress accumulates, and modifications take place. Earthquakes, volcanic activity, and sluggish crustal motions are all possible outcomes of this phenomenon.
- Active Isostasy: Any locations that are undergoing quick changes, such as regions that are tectonically active or regions that are experiencing intense glaciations or erosion, are continuously going through the process of Isostatic adjustments.

Applications of Isostasy

- Geological Surveys: By gaining a grasp of Isostasy, one can aid in the prediction of changes in land elevation, which is of utmost importance for geological surveys and the comprehension of crustal deformations.
- Oil and Mineral Exploration: In the field of exploratory geology, Isostatic balance models are employed to provide estimations on the potential locations of resources, taking into account differences in crustal density.
- Climate Studies: Isostatic adjustments, which include post-glacial rebound, offer valuable insights about the glacial covering of the past and contribute to the fields of study that investigate global climate change.

To summarize, Isostasy is the process that occurs when the crust of the Earth and the mantle of the Earth remain in a state of equilibrium. It provides an explanation of how the surface of the Earth reacts to different loads (such as mountains, glaciers, and sediments) and how it adjusts throughout time in order to keep the gravitational balance.

Chapter 24
Geosynclines

Introduction

The term "geosyncline" refers to a significant depression in the crust of the Earth that is rich in sedimentary and volcanic materials. Mountain ranges can be formed from these depressions, which are frequently connected with tectonic activity and have the potential to eventually become folded, faulted, and uplifted when this happens. The geosyncline theory was the dominant concept that geologists utilized in order to explain the formation of mountains prior to the development of plate tectonics. The length of geosynclines can range from hundreds to thousands of kilometers, and they are frequently connected with the boundary of tectonic plates. One of the earliest understandings of the forces and processes that shape mountain ranges and tectonic basins was offered by the concept of a geosyncline. As early as the middle of the 19th century, the American geologist James Hall and his colleague James Dwight Dana claimed that the Appalachian Mountains were formed by the folding and uplift of sediments in a long, subsiding trough. This was the first time that the term "geosyncline" was used.

Geosyncline Theory before Plate Tectonics

Mountain range growth was primarily explained by the geosyncline model prior to the present theory of plate tectonics. It proposed that the Earth's crust subsided, creating enormous, elongated basins that gradually filled with sediments and volcanic materials before being uplifted and folded by compressional forces to produce mountains. The mechanisms that led to folding and faulting in the Earth's crust, as well as the thick layers of marine deposits observed in mountain ranges, were both explained by the theory. The fundamental processes that resulted in sediment buildup, crustal subsidence, and the eventual uplift and deformation of mountain ranges, however, were not explained by the geosyncline model. A more thorough and mechanistic understanding of these processes was made possible by the development of plate tectonics in the 1960s, which essentially eclipsed the geosyncline model. However, in historical geology, the idea is still helpful in explaining sedimentary basins and mountain-building processes.

Types of Geosynclines

A. Ortho-geosyncline

- **Description**: Long, narrow depressions called ortho-geosynclines are usually seen inside mobile belts or at the margins of continental plates. Over millions of years, these areas deposit significant amounts of silt, creating long basins.
- **Sedimentation**: Ortho-geosynclines are filled with thick layers of sediments that are frequently obtained from surrounding landmasses. Detrital debris, shallow marine deposits, and deep-sea sediments could all be included in the sequence.
- **Tectonic Deformation**: Tectonic forces exert their influence on the sediment-filled basin over a period of time, causing the sediments to undergo deformations such as folding, faulting, and thrusting. The tectonic compression that occurs can result in the construction of Fold Mountains, which are formed when the crust of the Earth buckles and rises simultaneously.
- **Examples**: Ortho-geosynclines are commonly associated with the formation of major mountain chains such as the **Himalayas**, **Alps**, and **Rockies**. These mountain belts are formed due to the intense deformation of thick sedimentary layers deposited in the ortho-geosynclinal basins.

B. Eu-geosyncline

- **Description**: The geosynclinal troughs known as Eu-geosynclines are composed of a combination of sedimentary and volcanic materials inside of them. These regions are often located at a larger distance from stable cratonic regions, which are the cores of continents that have been stable for a long time, and they are more geologically active.
- **Volcanic Activity**: The eu-geosynclines, in contrast to the ortho-geosynclines, are connected with a large amount of volcanic activity. This volcanic activity is a contributor to the production of volcanic rocks within the sedimentary sequence. Examples of volcanic rocks include lava flows and volcanic ash products.

- **Tectonic Setting**: This type of geosyncline is frequently observed in areas where there is active plate convergence, such as subduction zones. These zones are characterized by the movement of oceanic plates beneath continental plates, which results in the formation of sedimentation and volcanism.
- **Examples**: The Andes mountain range in South America is an example of a region with eu-geosynclines, where volcanic rocks are interspersed with sedimentary layers, resulting from the subduction of the Nazca plate beneath the South American plate.

C. Mio-geosyncline

- **Description**: It is in the vicinity of cratonic zones that mio-geosynclines are formed. These are areas where sedimentary deposition takes place without major volcanic activity. The stability of these geosynclines is higher than that of eu-geosynclines, and they are situated in close proximity to continental shields or consistent regions of the Earth's crust.
- **Sedimentation**: The sedimentary layers in mio-geosynclines have a comparatively thinner thickness when compared to the thickness of sedimentary layers in other types of geosynclines. A significant portion of the sediments originate from the continents, with additional contributions coming from rivers, lakes, and shallow oceans.
- **Lack of Volcanism**: Mio-geosynclines, in contrast to eu-geosynclines, do not include any volcanic materials since they are situated in geographical areas that are not close to active tectonic plate borders or volcanic zones
- **Examples**: The Appalachian Basin in the eastern United States is an example of a mio-geosyncline, where sediments were deposited near the stable North American craton during the Paleozoic era.

D. Taphro-geosyncline

- **Description**: Taphro-geosynclines constitute depressions or basins that are primarily created as a result of faulting and rifting forces. Rift valleys and graben valleys are two more names for these extended depressions or valleys.
- **Formation**: The formation of taphro-geosynclines occurs when the crust of the Earth undergoes stretching and thinning, resulting in the formation of normal faults. A profound rift or valley is formed as the land that lies between these faults begins to subside.
- **Tectonic Activity**: These places are frequently linked to divergent plate borders or rift zones, which are areas where the crust of the Earth is being driven apart.
- **Examples**: The **East African Rift Valley** is a well-known example of a taphro-geosyncline, where tectonic forces are pulling the African continent apart, forming deep rift valleys.

E. **Para-geosyncline**

- **Description**: A para-geosyncline is a type of geosyncline that is formed within the cratonic region, as opposed to forming at the borders of the craton. Due to the fact that it grows within the old, stable portions of continents, it is a sort of geosyncline that is particularly stable in comparison to other types.
- **Sedimentation**: Para-geosynclines are characterized by the accumulation of thick sequences of sediments. However, unlike ortho- or eu-geosynclines, these sediments are not subjected to severe tectonic forces since they are not subjected to the same extent. When compared to other strata, the sedimentary layers are often relatively undisturbed.
- **Examples**: The **Williston Basin** in North America is an example of a para-geosyncline. It is located within the North American craton and has a thick accumulation of sedimentary rocks without significant tectonic deformation.

F. **Zeugo-geosyncline**

- **Description**: A particular kind of para-geosyncline that contains marginal uplifts is called a Zeugo-geosyncline. Tectonic pressures that raise the basin's boundaries frequently result in these uplifts, which are elevated regions along the geosyncline's edges.
- **Formation**: A more complex tectonic setting is created inside the basin as a result of the elevated areas that form at the edges of the geosyncline. This causes differential subsidence.
- **Examples**: Zeugo-geosynclines can be found in regions where cratonic basins are influenced by nearby tectonic activity, resulting in marginal uplifts.

G. **Auto-geosyncline**

- **Description**: A para-geosyncline, which is a form of geosyncline, is characterized by the absence of marginal uplifts. In the process of its development, it takes place within a cratonic region and is distinguished by the buildup of sediment along its boundaries, without any substantial tectonic deformation or uplift.
- **Stability**: Due to the fact that auto-geosynclines originate from stable cratonic zones and do not entail any marginal uplifts, they have a tendency to maintain their geological stability throughout extended instances of time.
- **Examples**: The Hudson Bay Basin in Canada is an example of an auto-geosyncline, where sedimentary layers accumulated within the stable Canadian Shield without significant tectonic disturbances.

Geologists can better understand how distinct sedimentary basins emerge and change throughout geological time, as well as how they contribute to the formation of mountain ranges and other tectonic features, by examining these diverse forms of geosynclines.

Every kind of geosyncline has a distinct function in the tectonic history and crustal dynamics of the Earth.

Evolution of a Geosyncline

Initial subsidence, sediment deposition, tectonic deformation, and eventual uplift are the crucial phases in the formation of a geosyncline. Tens to hundreds of millions of years pass between these stages.

Stage 1: Initial Subsidence

The Earth's crust is stretching downward, or subsiding, as part of the initial stage of geosyncline development. There are several tectonic mechanisms that can do this:

A. **Tectonic Extension**: There are instances where the lithosphere undergoes a thinned and subsiding process due to the stretching of the crust at rift zones or divergent boundaries. This process creates space for the deposition of sediment.

B. **Loading by Sediment and Volcanism**: The accumulation of sediments and volcanic deposits results in the underlying crust sinking further, resulting in the formation of a depression.

It is possible for the geosyncline to eventually stretch for thousands of kilometers, which will result in the formation of a vast basin that is elongated in shape.

Stage 2: Sediment Accumulation

The geosyncline will become an ideal area for sedimentation once it has been lowered to its previous level. There are a variety of surrounding landmasses, river systems, volcanic eruptions, and oceanic activities that can all be considered sources of sediment. In the course of time, the geosyncline may become filled with extensive sequences of sedimentary rocks, which may have a thickness of anywhere between 10 and 20 kilometers.

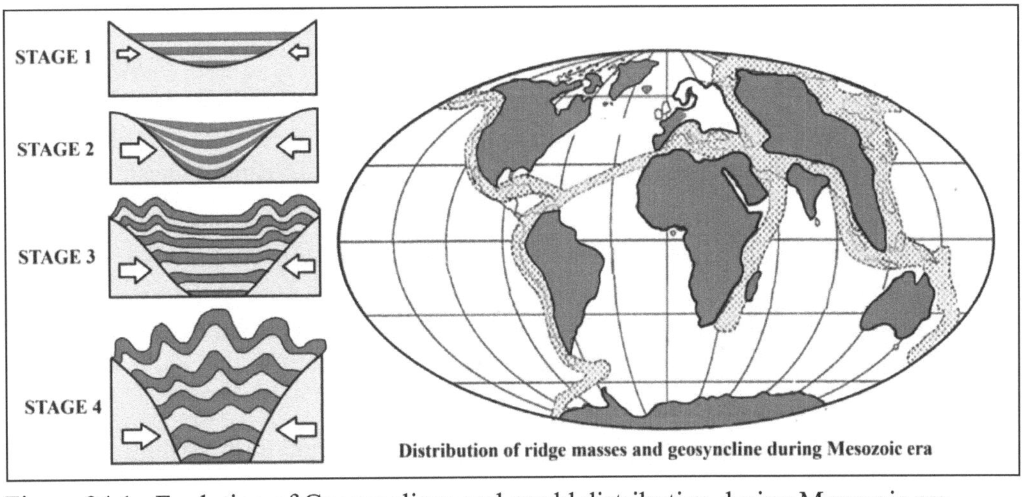

Figure 24.1: Evolution of Geosynclines and world distribution during Mesozoic era.

- Eugeosynclines are characterized by the presence of deep-sea sediments, which include muds, volcanic ash, and turbidites, among other types of elements.
- Miogeosynclines are characterized by the presence of shallow-water deposits, which include sandstones, limestones, and shales. These sediments are often produced in habitats that are either on the continental shelf or on slopes.

Sediments that have been accumulating in the geosyncline for an extended period of time are what prepare the region for tectonic deformation in the future.

Stage 3: Tectonic Deformation and Uplift

The sedimentary basin is deformed and uplifted during the third stage of geosyncline evolution. This happens frequently along convergent plate borders when tectonic pressures shorten and compress the Earth's crust. Mountain ranges are formed as the sediments in the geosyncline are forced upward, faulted, and folded as the plates meet.

The forces responsible for deformation include:

- **Compression**: The geosyncline experiences compressional stress from convergent tectonic plates, which causes the sedimentary strata to fold and fault.
- **Thrust Faulting**: This can result in the formation of mountain ranges that are made up of stacked layers of rock. Large parts of the crust can be pushed over one another.
- **Metamorphism**: The extreme pressures and temperatures that occur during the process of deformation have the potential to give rise to metamorphism in the sedimentary and volcanic rocks that arc locatcd in the geosyncline. This transformation can result in the production of schists, gneisses, and other metamorphic rocks.

Stage 4: Mountain Building (Orogeny)

The last step of geosyncline evolution is called orogeny, which refers to the formation of mountains. As a result of the ongoing crustal deformation, the geosyncline is being folded, elevated, and eventually changed into a mountain range configuration. The surface of the Earth undergoes enormous transformations as a result of this process, which can take tens of millions of years to complete.

- **Continental-Continental Collision**: The sediments are subjected to enormous compression and are pushed upward into towering mountain ranges if the geosyncline is situated at a continental-continental collision zone, as is the case with the Himalayas.
- **Oceanic-Continental Convergence**: It is possible for the geosyncline to be elevated as volcanic arcs and fold-and-thrust belts along oceanic-continental convergent borders, as is the case in the Andes.

Either the mountain range that has been elevated will continue to expand as a result of continuous tectonic action, or it will gradually disintegrate over the course of geological time.

Geosyncline Theory and Modern Plate Tectonics

For the most part, the geosyncline hypothesis has been declared outmoded by the contemporary theory of plate tectonics, despite the fact that it was effective at its time. The theory of plate tectonics offers a more all-encompassing framework for comprehending the processes of crustal deformation, sedimentation, and mountain creation.

Plate Tectonics and Sedimentation: Within the framework of plate tectonics, geosynclines are characterized by the presence of sedimentary basins situated near convergent boundaries. These basins are formed when one tectonic plate is subducted beneath another, hence generating a space for the accumulation of sediment.

Subduction Zones: It is currently understood that a significant number of geosynclines originated at subduction zones. These zones are characterized by the subduction of oceanic plates beneath continental plates, which ultimately results in the formation of deep marine basins and volcanic arcs.

Orogeny in Plate Tectonics: The process of orogeny, which refers to the formation of mountains, can be better understood by examining the interactions between tectonic plates, particularly at the location of convergent margins. Mountain ranges are formed as a result of the thickening, deformation, and uplift of the Earth's crust, which is caused by the collision and convergence of plates around the Earth.

Although the geosyncline model has been replaced by the plate tectonics theory, it continues to hold significant importance in the field of historical geology. This is especially true when it comes to defining sedimentary basins and the large-scale processes that have created ancient mountain ranges. The notion of geosynclines is an essential one in the field of geological history. It provides a model that can explain the deposition of sediments and the subsequent uplift and deformation of these deposits into mountain ranges. In spite of the fact that plate tectonics has essentially supplanted the geosyncline theory, it continues to offer a vital framework for comprehending sedimentary basins and the processes that result in the building of mountains. Geosynclines can be thought of as large, long-lasting troughs or basins that capture the history of tectonic and sedimentary processes. These processes have evolved over millions of years into the awe-inspiring mountain ranges that we see today. It is still extremely important to investigate these ancient basins, the sediments that they contain, and the tectonic forces that act upon them in order to gain an understanding of the geological history of the Earth as well as the processes that continue to shape itself.

Chapter 25
Origin of Oceans

Origin of Oceans

Extensive connections exist between the processes of plate tectonics and the formation of ocean basins as well as their subsequent evolution. Exploring the mechanics of continental rifting, seafloor spreading, and the dynamic interactions that occur at tectonic plate boundaries is neccssary in order to gain an understanding of the formation of oceans from a scientific standpoint.

1. Early Earth and Ocean Formation

A. **Primordial Earth**: When the Earth was first formed, it was a molten mass that gradually cooled and hardened to form a crust upon the surface. The release of water vapor and other gases into the atmosphere as a result of volcanic outgassing was a significant factor in the development of the seas today.

B. **Hydrosphere Formation**: As the Earth cooled, water vapor condensed into liquid water, which resulted in the accumulation of water in areas that were low-lying. Because of this condensation and buildup of water, the first oceans came into existence millions of years ago.

2. Continental Rifting and Ocean Basin Formation

A. **Continental Rifting**: The term "continental rifting" refers to the process by which tectonic forces cause a continental landmass to be torn apart. Rifting is results in the formation of a rift valley, which has the potential to eventually expand and deepen, so forming a new ocean basin. For instance, the East African Rift Valley is a contemporary illustration of the early stages of continental rifting on Earth.

B. Stages of Rifting:

a. Initial Stage: Faults and cracks in the continental crust are caused by the application of tensional forces.

b. Rift Valley Formation: As the process of extension continues, it results in the construction of a rift valley.

c. Linear Sea Formation: As the rift valley continues to develop, it may become flooded with seawater, resulting in the formation of a narrow sea.

d. Ocean Basin Formation: Continual spreading of the seabed results in the formation of a fully fledged ocean basin.

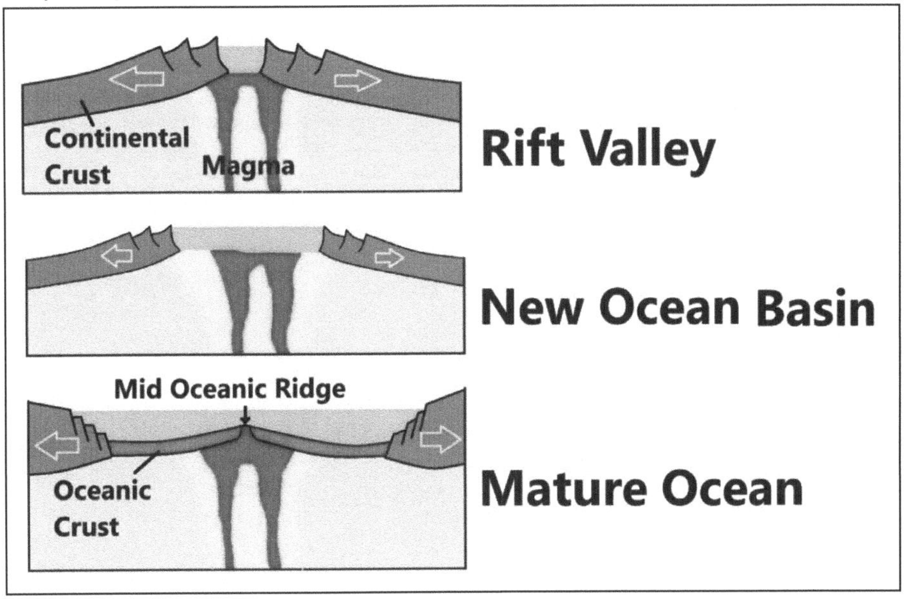

Figure 25.1: Figure illustrating Continental Rifting and Ocean Basin Formation

3. Seafloor Spreading

A. Mid-Ocean Ridges: Mid-ocean ridges are mountain ranges that are located underwater and are associated with the formation of new oceanic crust by the upwelling of magma from the mantle. In addition to having hydrothermal vents, these ridges are characterized by significant levels of volcanic activity. The Mid-Atlantic Ridge is an example of an active mid-ocean ridge that is contributing to the expansion of the Atlantic Ocean.

B. Mechanism of Seafloor Spreading: Magma has the ability to emerge from the mantle along divergent boundaries, resulting in the formation of new oceanic crust when it cools and solidifies. Convection from the mantle is the force that causes the freshly produced crust to flow away from the ridge axis in a symmetrical manner on both sides. There is evidence that the process of seafloor spreading is supported by magnetic anomalies and the age-dating of oceanic crust.

4. **Evolution of Ocean Basins**

A. **Mature Ocean Basins**: Ocean basins that have reached their full maturity are distinguished by the presence of well-developed mid-ocean ridges, abyssal plains, and deep-sea trenches. As an illustration, the Pacific Ocean is the largest and oldest ocean basin, and it was formed by vast subduction zones and seafloor spreading.

B. **Subduction and Ocean Basin Closure**: When oceanic crust is subducted beneath a continental or another oceanic plate, ocean basins have the potential to close because of this process. The creation of deep-sea trenches and volcanic arcs is the result of this process. As an illustration, the Himalayas were formed as a result of the collision between the Indian Plate and the Eurasian Plate, which occurred as a result of the closure of the ancient Tethys Ocean.

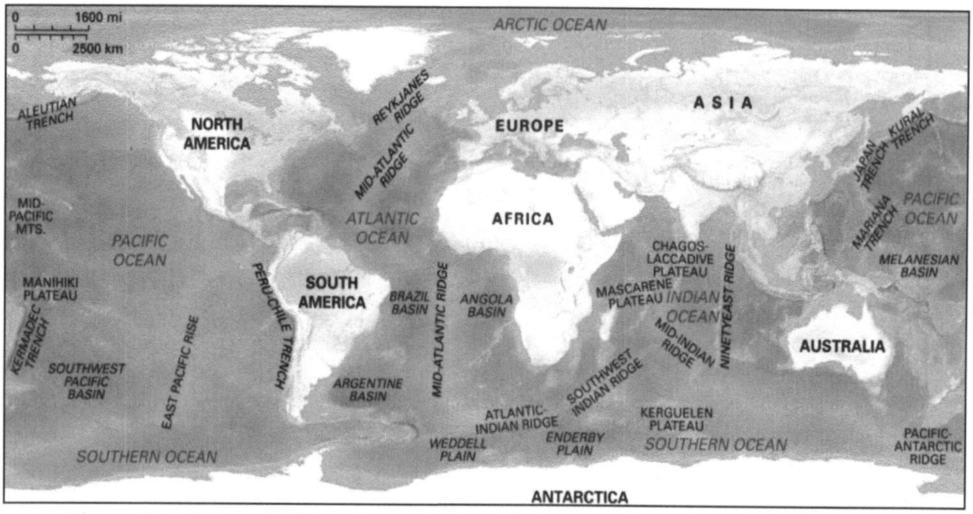

Figure 25.2: Map of Major Ocean Basins (Reference: Encyclopædia Britannica.)

5. **Characteristics of Ocean Basins**

A. **Abyssal Plains**: Abyssal plains are sections of the deep ocean floor that are flat and stretch away from mid-ocean ridges when they are located. The buildup of fine sediments on top of the basaltic oceanic crust is what gives rise to these formations today. For example, the broad, level regions that makes up the floor of the Atlantic Ocean.

B. **Oceanic Trenches**: Oceanic trenches are deep depressions that are limited in width and produced in subduction zones, which are areas where one plate is thrust beneath another plate. Trenches are the locations in the ocean that are the deepest and are associated with significant levels of seismic activity. As an illustration, the Mariana Trench, this is the deepest trench in the entire planet.

C. **Mid-Ocean Ridges**: Mid-ocean ridges are mountain ranges that are continuous and located underwater. They are used to indicate the boundaries of divergent plates. Significant volcanic and hydrothermal activity can be found at these locations, which are also sites of the production of new oceanic crust. Consider the Mid-Atlantic Ridge as an example.

6. Evidence for Ocean Basin Formation

A. **Magnetic Anomalies**: The reversals of the Earth's magnetic field are recorded by symmetrical patterns of magnetic stripes that are located on either side of mid-ocean ridge sites. The existence of these stripes is proof that the marine crust has been continuously forming and moving outward.

B. **Age of Oceanic Crust**: Samples of oceanic crust that have been subjected to radiometric dating have revealed that the age of the crust increases with increasing distance from the mid-ocean ridges. At the axis of the ridge, the crust that is the youngest may be discovered, while the crust that is the oldest can be found the farthest away.

C. **Heat Flow Measurements**: It is possible that active magma upwells and crust buildings are occurring at mid-ocean ridges because of the higher heat flux. This is compatible with the cooling and aging of the oceanic crust, which causes the flow of heat to decrease as one move further away from the ridge.

D. **Seismic Activity**: The mid-ocean ridges, transform faults, and subduction zones are the specific locations where earthquakes are most likely to occur. There is a correlation between the distribution of seismic activity and the tectonic processes that are involved in the development of ocean basins.

Plate tectonics is the driving force behind the complicated process that gives rise to oceans. The process of continental rifting is responsible for the beginning of the construction of new ocean basins, which are then further developed by the spreading of seafloor at mid-ocean ridges. The evolution of ocean basins and the traits they exhibit are ultimately shaped by the dynamic interactions that occur at divergent and convergent boundaries. The geological history of the Earth's surface and the dynamics of the surface at the present time can be better understood by gaining an understanding of these processes.

Chapter 26
The origin of continents

There are many different geological processes and ideas that are involved in the origin of continents, making it a complicated issue in geology. Regarding the topic, the following are some comprehensive notes:

1. Theories on the Origin of Continents

A. Continental Drift Theory: In 1912, Alfred Wegener put out his proposal. There was once a single landmass known as Pangaea, which included all of the continents. This landmass began to disintegrate approximately 200 million years ago.

Evidence:

a. Fit of the Continents: It seems as though the coastlines of continents like South America and Africa are like to pieces of a jigsaw puzzle, fitting together in a coordinated manner.

b. Fossil Evidence: Comparable fossils of plants and animals have been discovered on continents that are currently much separated from one another.

c. Geological Evidence: The rock formations and mountain ranges that are on different continents that is similar to one another.

d. Paleoclimatic Evidence: Evidence of former climates, such as glacial deposits, discovered in locations that are currently classified as tropical.

B. **Plate Tectonics Theory:** Plate Tectonics Theorydeveloped in the 1960s and 1970s. There are a number of major and tiny tectonic plates that make up the lithosphere of the Earth. These plates exist on top of the semi-fluid asthenosphere that lies beneath them.

Mechanisms:

a. Sea-Floor Spreading: At mid-ocean ridges, new oceanic crust is formed and then spreads outward, causing continents to become farther distant from one another.
b. Subduction Zones: The subduction zones are responsible for the destruction of the old oceanic crust, as it sinks back down into the mantle.
c. Transform Faults: Horizontally, plates slide past one another as they pass.

2. Geological Processes in the Formation of Continents

A. Craton Formation: Ancient and stable components of the continental lithosphere are referred to as cratons.

Processes:

a. Accretion: Smaller landmasses or terranes come into contact with a growing craton and attach themselves to it.
b. Metamorphism and Magmatism: Metamorphism and magmatic activity are both caused by heat and pressure, which ultimately results in the stabilization of the craton.

B. Orogeny (Mountain Building): At the limits of convergent plate boundaries, when tectonic plates collide with one another.

Processes:

a. Subduction: The process of one plate being pushed beneath another result in the formation of mountain ranges and volcanic arcs.
b. Collision: The collision of two continental plates results in the formation of massive mountain ranges such as the Himalayas.

C. Rifting: When a continental plate is stretched and thinned, this phenomenon takes place.

Processes:

Continental Rifting: The lithosphere undergoes a process of disintegration, resulting in the formation of rift valleys, which may eventually result in the formation of new ocean basins.

3. Evidence Supporting Plate Tectonics

A. Paleomagnetism: The magnetic characteristics of rocks are the individual of this study.

Key Evidence:

a. Magnetic Stripes: Recording reversals of the Earth's magnetic field are magnetic stripes that are symmetrical and found on the ocean floor.

b. Apparent Polar Wander: Indicating the movement of continents, paths that demonstrate the historical movement of the magnetic poles are presented.

B. **Distribution of Earthquakes and Volcanoes:** Plate borders are the locations where earthquakes and volcanoes are most likely to occur.

C. **Age of Oceanic Crust:** At mid-ocean ridges, the crust of the ocean is the youngest, whereas along subduction zones, it is the oldest.

4. Modern Technology in Studying Continental Origins

A. Satellite Geodesy: Utilizing global positioning system technology, it is possible to accurately measure the movement of tectonic plates.

B. Seismology: Investigates seismic waves in order to gain an understanding of the structure of the interior of the Earth as well as the characteristics of plate movements.

C. Geochronology: Methods of radiometric dating are utilized in order to ascertain the ages of rocks and the sequence of events that occurred in geological time.

5. Implications of Continental Movements

A. Climate Change: The movement of continents has an effect on the patterns of climate and the currents in the ocean.

B. Biodiversity: Continental drift and plate tectonics are two significant factors that have an impact on the evolution and spread of organisms.

C. Natural Hazards: There is a clear connection between tectonic activity and natural disasters such as earthquakes, volcanic eruptions, and tsunamis.

6. Historical Development of Theories

A. Early Observations

- Abraham Ortelius (1596): It has been proposed that continents were previously connected together, but they have since separated.
- Antonio Snider-Pellegrini (1858): In this illustration, the match of continents and similar fossil distributions was demonstrated.

B. Acceptance of Plate Tectonics

- Harry Hess (1962): Proposed the theory of sea-floor spreading.
- Vine and Matthews (1963): Provided paleomagnetic evidence for sea-floor spreading.
- John Tuzo Wilson (1965): Introduced the concept of transform faults and hotspots.

An extending and dynamic process that is impacted by a variety of geological mechanisms is the process by which continents first came into existence. To appreciate the history of the Earth and to forecast the occurrence of future geological events, it is essential to have a solid understanding of these processes.

Chapter 27
The origin of Mountains

The formation of mountains, commonly referred to as orogeny, is a complicated geological process that involves the interaction of tectonic plates and a number of other geological forces. I have included some thorough notes on the topic below:

1. Theories and Processes in Mountain Formation

A. Plate Tectonics Theory

a. **Key Concept:** The movement and interaction of the lithospheric plates of the Earth are the primary factors that result in the formation of mountains.

b. **Types of Plate Boundaries Involved:**

- Convergent Boundaries: When two plates come into contact with each other.
- Divergent Boundaries: As two plates move apart from one another.
- Transform Boundaries: A situation in which two plates move past each other in a horizontal line.

B. Convergent Boundaries and Orogeny

1. **Continental-Continental Convergence**: At the time of a collision between two continental plates, neither of the plates gets subducted because of their buoyancy.As a consequence of the collision, the crust undergoes compression, folding, and thickening. Example: The Himalayas, formed by the collision of the Indian Plate and the Eurasian Plate.

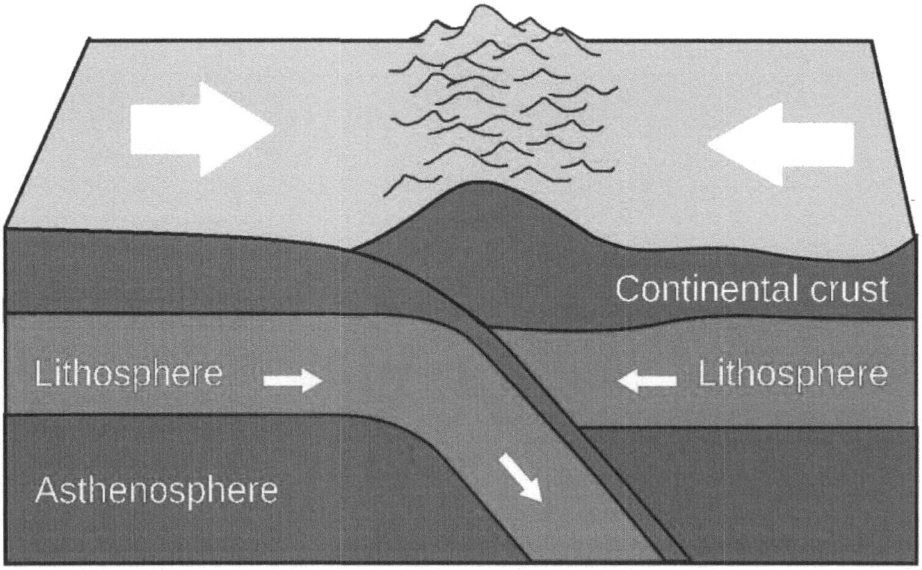

Figure 27.1: Figure showing Continental-Continental Convergence

2. **Oceanic-Continental Convergence**: A continental plate is able to subduct beneath an oceanic plate because of the higher density of the oceanic plate. The process of subduction is responsible for the creation of a volcanic mountain range along the margin of the continent. Example: The Andes Mountains, formed by the subduction of the Nazca Plate beneath the South American Plate.

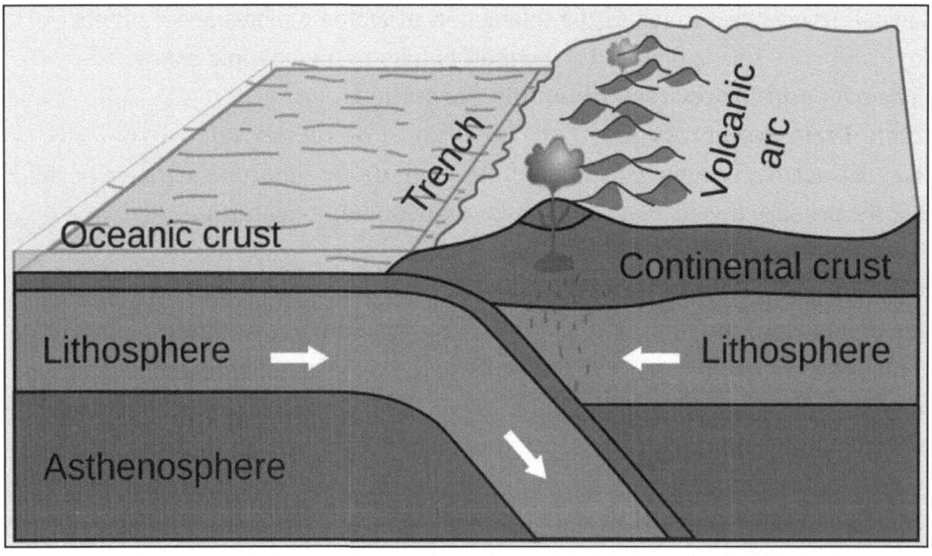

Figure 27.2: Figure showing Oceanic-Continental Convergence

3. **Oceanic-Oceanic Convergence:** There is a subduction of one of the oceanic plates beneath the other when two oceanic plates converge. A volcanic island arc is formed

as a result of this subduction, which has the potential to expand into mountainous regions. Example: The Japanese Archipelago.

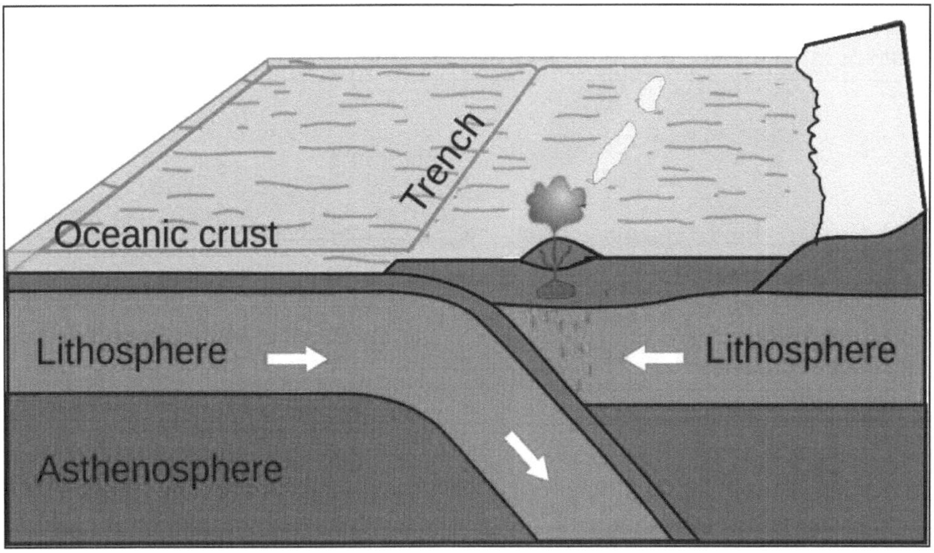

Figure 27.3: Figure showing Oceanic-Oceanic Convergence

C. **Divergent Boundaries (Rift Mountains):** Rift valleys, which can be formed when plates move apart and create rift valleys, are capable of elevating mountain ranges, such as the East African Rift.

D. **Transform Boundaries** (Fault Block Mountains): This phenomenon is characterized by the horizontal displacement along fault lines, such as the San Andreas Fault.

2. Types of Mountains and Their Formation

A. Fold Mountains

- Formation: A consequence of the collision of tectonic plates and the subsequent folding of those plates.
- Examples: Himalayas, Alps, Rockies, Andes.
- Processes:

a. Compression: The crust is compressed by horizontal forces, which results in the crust buckling and folding.

b. Anticlines and Synclines: Anticlines, which are upward folds, and synclines, which are downward folds, are the folds that generate the recognizable ridges and valleys.

B. Fault-Block Mountains

- Formation: The fracture and tilting of the Earth's crust along faults is what causes the formation of faults.
- Examples: Sierra Nevada, Harz Mountains.
- Processes:

a. Tensional Forces: In order to cause blocks to move up or down along fault lines, it is necessary to stretch the crust.
b. Horsts and Grabens: In order to form the mountain and valley structures, the blocks that are lifted (horsts) and the blocks that are dropped down (grabens) are used.

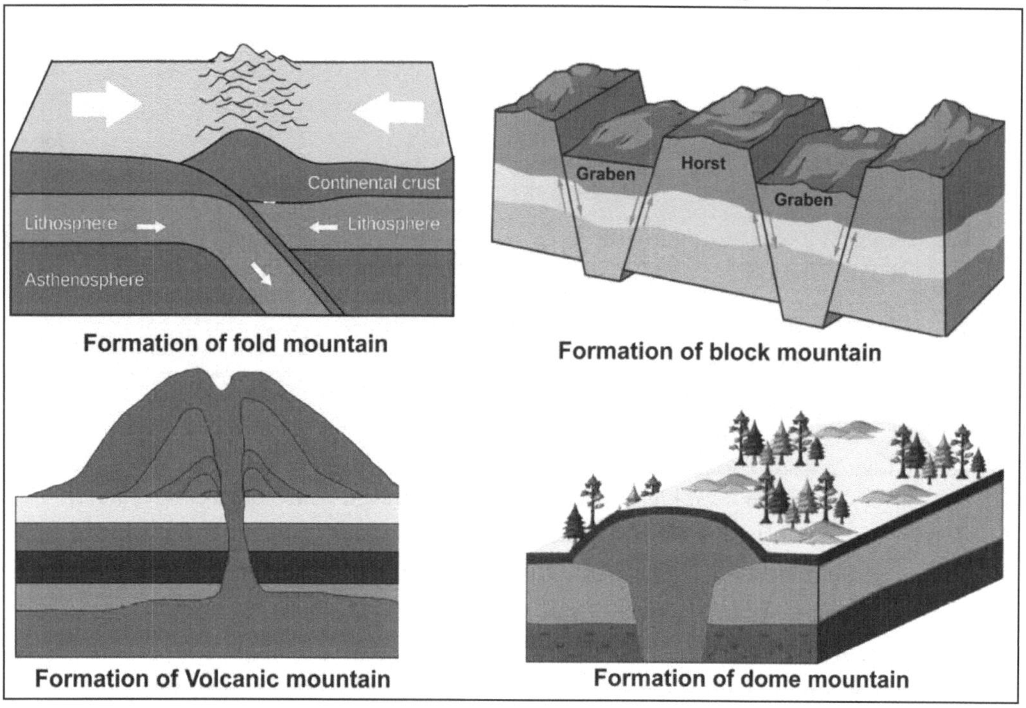

Figure 27.4: Types of Mountains

C. Volcanic Mountains

- Formation: The result of volcanic activity, which occurs when magma from deep under the Earth's crust becomes exposed to the surface.
- Examples: Mount Fuji, Mount St. Helens, Mauna Kea.
- Processes:

a. Eruption of Lava: A mountain is formed by the accumulation of successive layers of volcanic rock.
b. Volcanic Cones: An accumulation of volcanic material surrounding the vent results in the formation of a mountain that is formed like a cone.

D. Dome Mountains

- Formation: Produced by the intrusion of magma, which causes the rock layers that are on top of it to be pushed into the shape of a dome
- Examples: Black Hills, Adirondacks.
- Processes:

a. Magma Intrusion: Despite the fact that magma does not reach the surface, it causes the rock that is underneath it to swell.
b. Erosion: Over the course of time, erosion exposes the magma that has hardened and the rock that has been raised within it.

3. Geological Evidence and Processes

A. Stratigraphy and Rock Formations

- The study of rock layers and the structures of those layers can reveal information about the development of mountains.
- Key Evidence:

a. Folded and Faulted Layers: Indicating compressive forces is essential.
b. Metamorphic Rocks: Recommend the condition of deep burial and high pressure situations.
c. Igneous Intrusions: Volcanic and plutonic activities should be revealed.

B. Fossil Evidence: The presence of fossils in mountainous places can provide insight into the ancient conditions that existed and the movement of tectonic plates. Example: Marine fossils found at high altitudes indicate that the area was once under the sea before being uplifted.

4. Modern Tools and Techniques in Studying Mountain Formation

A. Satellite Geodesy: Measures the precise movements of tectonic plates using GPS technology.

Key Applications:

a. Monitoring Plate Movements: Provides assistance in comprehending the rates and directions of plate movements.
b. Detecting Uplift and Subsidence: Provides information regarding the vertical movements of the crust of the Earth.

B. Seismology: Seismic waves are studied in order to gain an understanding of the structure of the interior of the Earth as well as the dynamics of plate tectonics.

Key Applications:

a. Mapping Fault Lines: Identifies active and inactive faults.
b. Understanding Earthquakes: This study examines the relationship between earthquakes and the mechanisms that lead to the formation of mountains.

C. Radiometric Dating: The use of isotopic technologies allows for the determination of the ages of rocks and geological events.

Key Applications:

a. Dating Orogenic Events: It provides timings for the episodes that include mountain-building.
b. Correlating Rock Layers: Assists in determining the order in which geological occurrences occur.

5. **Historical Development of Theories**

A. **Early Observations**

a. James Hutton (18th Century): Developed the idea of uniformitarianism, which postulated that geological processes that are currently being observed may be traced back to the beginning of Earth's age.

c. Charles Lyell (19th Century): The concepts of Hutton were expanded upon and evidence was offered for the steady shift in geological conditions.

B. Modern Plate Tectonics

a. Alfred Wegener (1912): The hypothesis of continental drift, which laid the groundwork for the contemporary understanding of plate tectonics, was established.

b. Mid-20th Century Advances: Plate tectonics was supported by the discovery of paleomagnetism and sea-floor spreading, which gave substantial proof for the theory.

6. **Implications of Mountain Formation**

A. **Climate and Weather Patterns**

a. Mountains influence climate by affecting wind patterns and precipitation.

b. Examples:

- Rain Shadows: Mountains can block moisture, creating dry areas on their leeward sides.
- Orographic Lift: Mountains force air to rise, cool, and condense, leading to increased precipitation on windward slopes.

B. **Biodiversity and Ecosystems**

a. Mountains produce a wide variety of ecosystems and microclimates, which in turn support a wide variety of flora and fauna forms.

b. Examples:

- Endemic Species: Species that is unique to mountain regions.
- Vertical Zonation: Different ecological zones based on altitude.

C. **Natural Hazards**

a. A number of natural disasters, including earthquakes, landslides, and volcanic eruptions, are more likely to occur close to mountains.

b. Examples:

- Earthquakes: Common in tectonically active mountain regions.
- Landslides: Triggered by steep slopes, heavy rainfall, and seismic activity.

A dynamic and varied process, the origin of mountains is driven by the movement and interaction of tectonic plates. Mountains are formed as a result of this process. For the purpose of knowing the geological history of the Earth, anticipating future geological events, and managing the natural resources and risks connected with mountainous regions, it is vital to have a solid understanding of these processes.

Chapter 28
Concept and Application of Paleomagnetism

1. **Introduction:** The branch of research known as paleomagnetism examines the record of the Earth's magnetic field that has been retained in rocks, sediments, and other archeological materials. In addition to providing evidence that is essential to the ideas of plate tectonics and continental drift, this area of research also provides insights into the historical dynamics of the magnetic field of the Earth.
2. **Concept of Paleomagnetism**

A. Earth's Magnetic Field

a. Nature of the Magnetic Field: The magnetic field of the Earth is similar to that of a dipole magnet, which has a magnetic pole in the north and a magnetic pole in the south.

b. Generation: Produced by the motion of molten iron alloys in the outer core of the Earth, this is a process that is referred to as the geodynamo.

c. Magnetic Poles: It is important to note that the magnetic poles do not match exactly with the geographic poles and that they are prone to movement.

B. Remanent Magnetization

a. Thermoremanent Magnetization (TRM): Acquired by igneous rocks as they cool past the Curie temperature, which bring the magnetic minerals (such as magnetite) into alignment with the magnetic field of the Earth.

b. Detrital Remanent Magnetization (DRM): The magnetic particles are aligned with the magnetic field of the Earth as they are acquired by sediments as they settle through water.
c. Chemical Remanent Magnetization (CRM): Acquired by the process of chemical reactions, such as the production of new magnetic minerals, which align along with the magnetic field that is already present.

C. Magnetic Polarity Reversals

a. Geomagnetic Reversals: Throughout the course of geological history, the magnetic field of the Earth has undergone numerous instances of reversing its polarity, thereby reversing the positions of the magnetic north pole and the magnetic south pole.
b. Chronology: These reversals are documented in rock sequences, which together provide a history that is referred to as the geomagnetic polarity time scale (GPTS).

3. Techniques in Paleomagnetism

A. Sampling and Field Methods

a. Core Sampling: The drilling of cylindrical samples from rock formations is the first step.
b. Oriented Samples: In order to record the direction and inclination of the magnetic field at the time of rock formation, samples are properly orientated in order to provide this information.

B. Laboratory Analysis

a. Demagnetization: Isolating the principal remanent magnetization can be accomplished by the utilization of methods such as thermal and alternating field (AF) demagnetization.
b. Magnetometers: For the purpose of determining the magnetic characteristics of samples, various instruments, such spinning magnetometers and superconducting quantum interference devices (SQUIDs), are utilized.

C. Data Interpretation

a. Magnetic Declination and Inclination: An examination of the different angles that magnetic vectors have in relation to the horizontal plane (inclination) and the geographic north (declination).
b. Paleopoles: The positions of ancient magnetic poles that can be inferred from the remanent magnetization of rocks, which provides evidence for the movement of tectonic plates throughout history.

4. Applications of Paleomagnetism

A. Plate Tectonics and Continental Drift

a. Reconstruction of Past Positions: Paleomagnetic data gives scientists the ability to recreate the placements of continents in the past as well as the movement of tectonic plates across the earth.

b. Evidence for Continental Drift: The hypothesis of continental drift is supported by the fact that paleomagnetic data from separate continents can be related to one another.

B. Geomagnetic Polarity Time Scale

a. Chronostratigraphy: The GPTS is utilized as a dating tool in order to correlate and date sedimentary and geological sequences using the technology.

b. Magnetostratigraphy: The study of the magnetic characteristics of rock strata in order to ascertain the age of the rock layers and the correlation between them and the GPTS.

C. Volcanic and Seismic Activity

a. Volcanology: When it comes to understanding the history of volcanic eruptions and the development of volcanic regions, paleomagnetic studies are quite helpful.

b. Seismology: The examination of remanent magnetization can yield information regarding the seismic and tectonic activity that occurred in the past.

D. Archaeological Applications (Archaeomagnetism): The investigation of the magnetic characteristics of archaeological materials (such as bricks and burnt clay, for example) in order to ascertain the age of human artifacts and constructions and establish their purpose.

E. Environmental and Climate Studies

a. Paleoclimatology: In order to gain a better knowledge of the climate changes and environmental circumstances that occurred in the past, paleomagnetic data from lake sediments and ice cores are utilized.

b. Earth's Magnetic Field Variations: A better knowledge of how the Earth's magnetic field interacts with solar radiation and cosmic rays can be gained via the study of variations in the Earth's magnetic field throughout time.

5. Case Studies in Paleomagnetism

A. Mid-Atlantic Ridge (Sea-Floor Spreading): The formation of new crust at mid-ocean ridges and the symmetric pattern of magnetic reversals are both supported by the findings of paleomagnetic studies conducted on basaltic rocks that are located on the ocean floor.

B. Himalayas(Continental Collision): Paleomagnetic data derived from rocks in the Himalayan region provide evidence of the collision between the Indian and Eurasian plates, which led to the subsequent elevation of the mountain range.

C. North American Cordillera (Terrane Accretion): The hypothesis that multiple terranes, also known as crustal fragments, have been accreted to the North American continent as a result of plate tectonic processes appears to be supported by data from paleomagnetic analysis.

6. **Challenges and Future Directions**

A. **Diagenetic Alteration (Chemical Changes):**It is possible for diagenesis to change the magnetic properties of rocks in their initial state, which can make the interpretation of paleomagnetic data more difficult.

B. **Secular Variation (Short-Term Changes):** When doing paleomagnetic research, it is necessary to take into consideration certain short-term, secular changes that occur in the Earth's magnetic field.

C. **Technological Advancements**

a. Improved Instruments: The accuracy and resolution of paleomagnetic investigations are always being improved thanks to the advancements that have been made in magnetometers and data methods.

b. Interdisciplinary Research: When paleomagnetism is investigated in conjunction with other disciplines, such as geochronology, geochemistry, and structural geology, the scope of its applications and the depth of its comprehension will be expanded.

Paleomagnetism is an important subfield of geoscience that has made substantial contributions to our understanding of the magnetic field of the Earth, plate tectonics, and the history of geological formations. Scientists are able to recreate the migrations of continents in the past, date geological events, and obtain insights into the dynamic processes that occur on Earth by investigating the magnetic characteristics of rocks and sediments. In the future, the applications and significancc of paleomagnetism will continue to rise, bringing fresh perspectives on the history and evolution of our planet. This will be made possible by the advancement of technology and the expansion of research that spans multiple disciplines.

Chapter 29
Hypsography and Topography

1. **Introduction of Hypsography and Topography:** In the field of study that focuses on the surface of the Earth, hypsography and topography are two essential ideas. In addition to providing crucial data for a variety of scientific, engineering, and environmental applications, these phrases provide a description of the land's features and elevations.
2. **Definition of Hypsography**

A. **Concept of Hypsography**: The measurement and representation of the elevation and depth of land surfaces in relation to sea level is made possible through the practice of hypsography. Hypsos means "height," and graphia means "writing or description." These two Greek terms are where the phrase originates from.

B. **Hypsometric Data**

a. Elevation Measurements: Data points indicating the height of the land above sea level.

b. Depth Measurements: Data points indicating the depth of land below sea level, often pertaining to underwater topography.

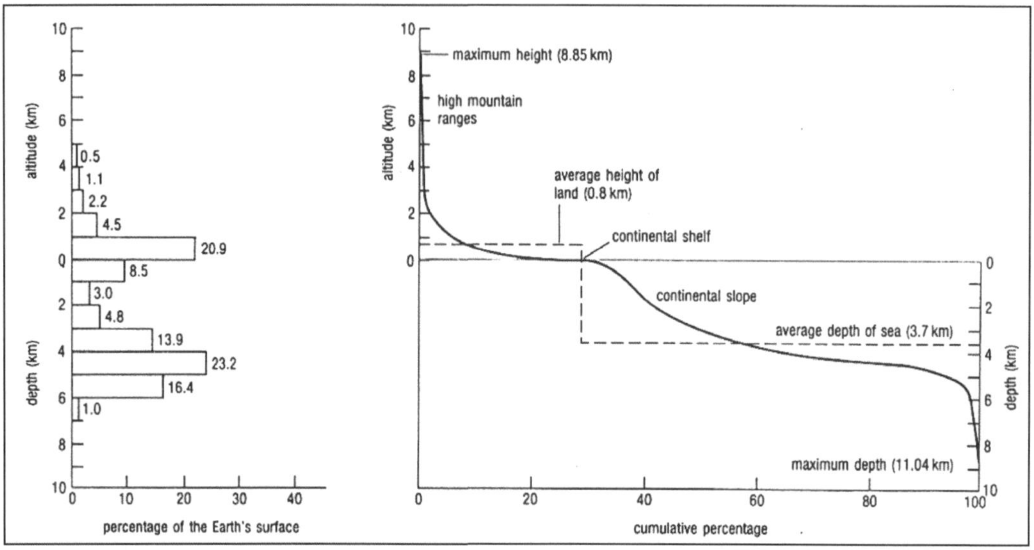

Figure 29.1: Hypsography of the Earth. Notice that Earth has two peaks in elevation, one for the continents, and the other for the ocean floors. (Reference: Wikipedia, the Free Encyclopedia)

C. Hypsometric Curve

a. Definition: A graphical representation that shows the distribution of elevations across a given area.

b. Uses:

- **Topographic Analysis:** The elevation profile of a region can be better understood with the help of this.
- **Erosion and Deposition Studies:** Indicates areas of erosion and sediment deposition.

D. Hypsographic Mapping

a. Contour Lines: Lines on a map that connect points of equal elevation, also known as isohypses.

b. Digital Elevation Models (DEMs): 3D representations of terrain elevations created from hypsometric data.

3. Definition of Topography

A. Concept of Topography: Topography is the study and representation of the physical properties and shape of the Earth's surface, including both natural and man-made features. Topography is a fundamental field of study. The term comes from the Greek words "topos" (place) and "graphia".

B. Topographic Features

a. Natural Features: Mountains, valleys, plains, rivers, lakes, and other landforms.

b. Man-Made Features: Roads, buildings, dams, and other infrastructure.

C. Topographic Maps

a. **Purpose:** The purpose of this is to depict the three-dimensional elements of the terrain on a map that is just two-dimensional.

b. **Elements:**

- **Contour Lines:** Indicate elevation and the shape of the land.
- **Symbols and Labels:** Represent various physical and cultural features.

c. **Scale:** The relationship between the distances on the map and the actual distances on the ground is expressed through this indicator.

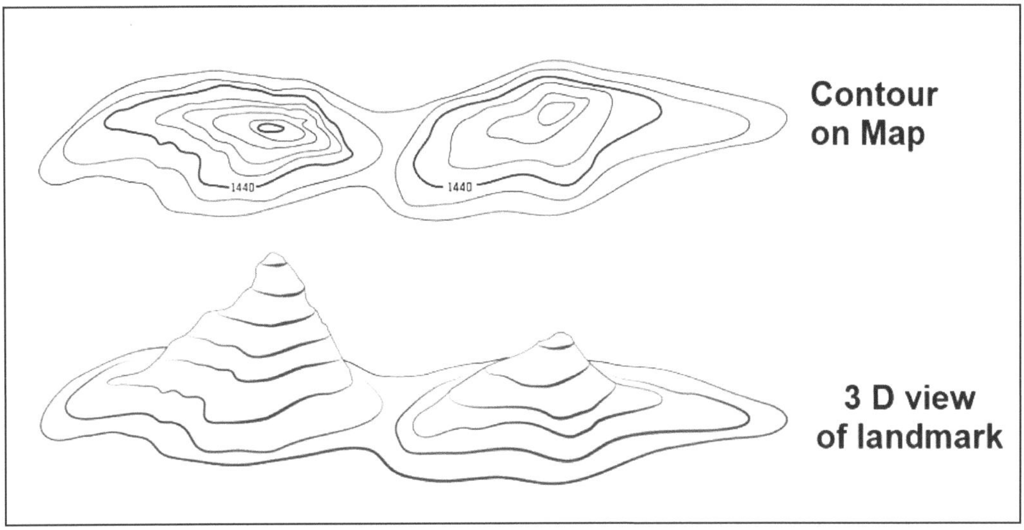

Figure 29.2: Contour lines showing the shape of the terrain

4. Techniques and Tools for Hypsographic and Topographic Studies

A. Surveying

a. Traditional Methods: The use of instruments such as theodolites, levels, and measuring tapes is the first step.

b. Modern Methods: The utilization of global positioning system technology to obtain accurate maps of elevation and location.

B. Remote Sensing

a. Satellite Imagery: Captures topography features on a vast scale during space exploration.

b. LiDAR (Light Detection and Ranging): Laser pulses are utilized in order to generate topographic maps with a high resolution.

C. Geographic Information Systems (GIS)

a. Data Integration: Combines data from a variety of sources in order to generate topographic and hypsographic maps that are extremely detailed.

b. Analysis Tools: The capability to do spatial analysis, modeling, and visualization of terrain features is provided.

5. Applications of Hypsography and Topography

A. Environmental Studies

a. Habitat Mapping: The first step is to have an understanding of the distribution of various ecosystems based on the topography and elevation.

b. Climate Studies: The study of the ways in which topography influences the climatic patterns of local and regional areas.

B. Urban Planning and Civil Engineering

a. Site Selection: The process of selecting suitable places for construction projects by using topographic data as a guide.

b. Infrastructure Design: When planning roads, bridges, and drainage systems, it is important to take into account the topography.

C. Disaster Management

a. Flood Modeling: By utilizing elevation data, flood zones can be predicted and flood defenses can be designed.

b. Landslide Risk Assessment: Locating locations that are prone to landslides by analyzing slope and different aspects of the terrain.

D. Agriculture and Forestry

a. Land Use Planning: The process of identifying regions that are appropriate for conservation, agriculture, and forestry.

b. Soil Erosion Control: To reducc soil crosion in hilly and mountainous places, the use of measures is being considered.

6. Case Studies in Hypsography and Topography

A. The Himalayas

a. Hypsographic Analysis: In order to gain a better understanding of the development and dynamics of the mountain range, one must first study the distribution of altitudes.

b. Topographic Mapping: Detailed maps that are utilized for activities such as hiking, scientific study, and emergency management.

B. Coastal Areas

a. Sea-Level Rise: Monitor and model the effects of rising sea levels on the topography of coastal areas.

b. Hypsometric Studies: Assessing the susceptibility of coastal areas to erosion and floods is the first step.

C. Urban Areas

a. Topographic Surveys: Conducting extensive surveys for urban development and infrastructure projects.

b. GIS Applications: The utilization of geographic information systems (GIS) for the purpose of planning and managing urban growth, transportation networks, and utility systems.

7. Challenges and Future Directions

A. Data Accuracy and Resolution

a. Challenges: Ensuring the precision and resolution of topographic and hypsographic data, particularly in locations that are difficult to access or located in regions that are distant.

b. Improvements: The quality of the data continues to improve as a result of developments in remote sensing and survey technology.

B. Climate Change Impacts

a. Challenges: Understanding and anticipating the effects of climate change on topography, such as the retreat of glaciers and the rise in sea level, is the first step.

b. Research Needs: Monitoring and modeling should be continued in order to provide information for adaptation and mitigation actions.

C. Technological Advancements

a. Emerging Tools: The creation is more sophisticated technologies for the collecting, processing, and visualization of data.

b. Interdisciplinary Approaches: By combining the findings of topographic and hypsographic research with those of other scientific fields, a more comprehensive understanding of Earth's systems can be achieved.

The disciplines of hypsography and topography are both extremely important areas of research because they offer valuable insights on the elevations and features of the Earth's surface. The fields of environmental studies, urban planning, disaster management, and a great number of other fields can all benefit from the widespread use of these disciplines. The expansion of our understanding and application of hypsographic and topographic data will continue to be facilitated by technological advancements and research that draws from a variety of disciplines. This will assist in addressing the difficulties and possibilities that will be present in the future.

Chapter 30
Principle of Uniformitarianism

1. **Introduction to the Principle of Uniformitarianism:**
   The uniformitarianism principle is a fundamental idea in geology that proposes that the features of the Earth were formed by processes that were continuous and consistent over extended periods of time. In many cases, it is summed up by the statement "the present is the key to the past." Within the context of the study of Earth's history, this chapter delves into the origins of uniformitarianism, as well as its fundamental concepts and applications.
2. **Historical Development of Uniformitarianism**

A. Early Geological Theories

a. Catastrophism: The hypothesis that was prevalent just before the development of uniformitarianism, which proposed that the characteristics of the Earth were established by momentous, fleeting, and catastrophic occurrences.

b. Biblical Influence: The literal interpretations of biblical events, such as the Great Flood, were a significant source of inspiration for a great number of early hypotheses.

B. James Hutton and the Birth of Uniformitarianism

a. James Hutton (1726-1797): Hutton, who is sometimes regarded as the "father" of modern geology, claimed that the Earth was formed by processes that are still occurring today and are comparable to those that are observed today.

b. Publication of "Theory of the Earth" (1785): By rejecting the catastrophist viewpoint, Hutton's work created the foundations for uniformitarianism and established its foundation.

C. Charles Lyell and the Popularization of Uniformitarianism

a. Charles Lyell (1797-1875): Built upon Hutton's ideas and popularized them through his seminal work, "Principles of Geology" (1830-1833).

b. Lyell's Influence: Lyell's detailed observations and arguments solidified uniformitarianism as the dominant paradigm in geology.

3. Key Concepts of Uniformitarianism

A. Continuous and Gradual Processes

a. Erosion and Sedimentation: The processes of erosion, sedimentation, and volcanic activity are gradual processes that take place over the course of millions of years.

b. Plate Tectonics: The movement of tectonic plates, which is responsible for the formation of mountains and earthquakes, occurs over extremely long lengths of time.

B. Present Processes Reflect the Past

a. Geological Record: Scientists are able to deduce the history that is recorded in rock formations by observing the geological processes that are now occurring.

b. Fossil Evidence: It is possible to gain insight into the conditions of the environment in the past by analyzing the distribution and types of fossils.

C. Principle of Actualism

a. Definition: Actualism is strongly connected to uniformitarianism, which places an emphasis on the idea that natural laws and processes on the present day remain unchanged over the course of time.

b. Application: Geologists are able to use observations made in the current day to interpret geological events that occurred in the past thanks to this principle.

4. Evidence Supporting Uniformitarianism

A. Rock Formations and Stratigraphy

a. Stratigraphic Layers: The examination of sedimentary rock strata exposes a chronological sequence of events, which provides evidence for the slow deposition that occurs over the course of time.

b. Unconformities: The geological record has gaps that point to periods of erosion or non-deposition, which are compatible with processes that occur over a lengthy period of time.

B. Fossil Record

a. Succession of Life Forms: A constant progression and eventual extinction of species over the course of millions of years is demonstrated by the fossil record.

b. Paleoenvironmental Indicators: As a result of fossils, we may learn about the temperatures, sea levels, and ecological circumstances that existed in the past.

C. Radiometric Dating

a. Isotopic Decay: Radiometric dating techniques are used to determine the absolute ages of rocks and fossils by measuring the decay of radioactive isotopes.

b. Time Scales: The large time scales that are required for geological processes are confirmed by these approaches, which are consistent with the uniformitarianism theory.

5. Applications of Uniformitarianism in Geology

A. Geological Mapping and Field Studies

a. Rock Cycle: Geologists are able to more accurately forecast the creation and alteration of rocks over time when they have a solid understanding of the rock cycle.

b. Field Observations: In order to comprehend geological conditions and events that occurred in the past, geologists use observations from the current day.

B. Paleoenvironmental Reconstruction

a. Climate Change Studies: Scientists are able to gain a better understanding of the current trends in climate change and future projections by examining climates that existed in the past.

b. Sea-Level Changes: In order to shed light on the variables that are contributing to the current rise in sea levels, it is possible to conduct an investigation into the sea levels that have occurred in the past.

C. Natural Hazard Assessment

a. Earthquake Prediction: In order to identify potential threats in the future and to make preparations for them, it is important to conduct an analysis of seismic activity from the past.

b. Volcanic Eruptions: When it comes to anticipating and mitigating the effects of future eruptions, having a solid understanding of the volcanic activity that has occurred in the past is quite helpful.

6. Criticisms and Limitations of Uniformitarianism

A. Acknowledgment of Catastrophic Events

a. Modern Perspective: It is now acknowledged that catastrophic events, like as asteroid strikes and supervolcanic eruptions, have also played a substantial part in forming the Earth. This development comes despite the fact that uniformitarianism continues to be a fundamental principle in the field of geology.

b. Integrated Approach: The current state of geological knowledge incorporates both slow and catastrophic events.

B. Temporal and Spatial Scales

a. Limitations: It is necessary to exercise caution when interpreting the fact that certain processes may function at different rates or scales than those that are observable today.

b. Complex Systems: The systems that make up the Earth are intricate and have the potential to display non-linear and unpredictable responses throughout a variety of time ranges.

7. Case Studies in Uniformitarianism

A. The Grand Canyon

a. Stratigraphic Analysis: The rock layers that make up the Grand Canyon provide a precise record of the history of the Earth. These layers have been formed as a result of progressive sedimentation and erosion over the course of millions of years.

b. River Erosion: In the process of sculpting the canyon, the Colorado River played a significant part, which displays the force of gradual and continuous geological processes

B. Plate Tectonics and Mountain Building

a. Himalayas: The continuous collision between the Indian and Eurasian plates is a striking example of how modest tectonic movements over the course of millions of years can result in the formation of enormous mountain ranges.

b. Mid-Atlantic Ridge: The spreading of the seafloor at this divergent boundary is an example of how new crust is continuously being created.

8. Future Directions in the Study of Uniformitarianism

A. Technological Advancements

a. Remote Sensing: Satellite photography and remote sensing technology have advanced to the point that they can now provide extensive information on geological processes.

b. Geochronology: The development of more advanced dating methods continues to deepen our comprehension of the history of the Earth.

B. Interdisciplinary Research

a. Climate Science: Enhanced research into historical climatic shifts and the influence those shifts had on geological processes can be achieved by collaboration with climate scientists.

b. Biology and Ecology:The integration of paleontological data with ecological models is beneficial to the process of comprehending the interactions that occur between geological processes and all forms of life.

By placing an emphasis on the significance of processes that are both continuous and constant, the idea of uniformitarianism brought about a revolutionary change in our view of the history of the Earth. In spite of the fact that contemporary geology acknowledges the significance of catastrophic events, uniformitarianism continues to be an essential idea that serves as a foundation for the interpretation of geological records and the investigation of the dynamic systems of the Earth. It is through continual research and technology breakthroughs that uniformitarianism continues to provide significant insights into the history, the present, and the future of our planet.

Chapter 31
Geomorphic Agents

Introduction

The scientific study of landforms and the processes that create them is known as geomorphology. This area of study refers to the investigation of the dynamic interactions that take place between the surface of the Earth and a variety of geomorphic agents, such as water, wind, ice, and biological activity. For a complete comprehension of the formation and development of landscapes, it is vital to have an understanding of these agents and the processes that they drive.

Definition of Geomorphic Agents: The term "geomorphic agents" refers to natural forces or entities that are responsible for the processes of weathering, erosion, movement, and deposition of materials on the surface of the Earth.

Types of Geomorphic Agents

1. **Water:** Certain water geomorphic agents include precipitation, rivers, seas, oceans, and groundwater, among other things.
2. **Wind:** The geomorphic agents of wind are air currents that carry sediments.
3. **Ice:** Ice sheets and glaciers are examples of geomorphic agents.
4. **Biological Activity:** The behaviours of humans, animals, and plants are all examples of biological agents that contribute to geomorphic processes.

Water as a Geomorphic Agent

1. **Fluvial Processes**

A. River Systems: Rivers have a significant role in the processes of erosion, movement, and deposition.

B. Erosion: Hydraulic action, abrasion, solution, and attrition are some of the processes that come under the category of erosion.

C. Transportation: Different methods, including traction, saltation, suspension, and solution, are used to transport sediments.

D. Deposition: This phenomenon takes place when the river loses its energy, which results in the production of characteristics such as floodplains, alluvial fans, and deltas respectively.

2. **Coastal Processes**

A. Wave Action: Sediments are deposited, transported, and eroded along coasts with the help of waves.

B. Tides and Currents: Sediments are transported down the coast by tidal movements and ocean currents, which in turn create coastal landscapes.

C. Erosional Features: Stacks, sea arches, and cliffs are examples of erosional landforms.

D. Depositional Features: Barrier islands, spits, and beaches are examples of depositional features.

3. **Groundwater Processes (Karst Topography):** Groundwater processes, also known as Karst topography, are formed when soluble rocks, such as limestone, are dissolved by groundwater. This process results in the formation of land features such as caves, sinkholes, and subterranean streams.

4. **Glacial Processes**

A. Ice Movement: Movement of Ice Glacier is responsible for the formation of the landscape via their plucking and abrasion action.

B. Erosional Features: U-shaped valleys, fjords, and cirques belong to the category of erosional features.

C. Depositional Features: Moraines, drumlins, and eskers are components of the depositional features.

Wind as a Geomorphic Agent

1. **Aeolian Processes**

A. Erosion: Both deflations, which involve the removal of small particles, and abrasion, which has the effect of sandblasting, are types of erosion.

B. Transportation: Surface creep, Saltation, and suspension are the three methods that are used for transporting sediments.

C. Deposition: The formation of dunes, loess deposits, and sand sheets is referred to as deposition.

2. Desert Landscapes

A. Erosional Features: The erosive features of the desert include Yardangs, which are streamlined ridges, and desert pavements.

B. Depositional Features: The sand dunes, such as barchans and transverse dunes, as well as the loess plains, are examples of depositional features.

Ice as a Geomorphic Agent

1. Glacial Processes

1. Erosion: Glaciers are responsible for the erosion of the landscape via processes like as abrasion and plucking.
2. Transportation: The movement of glaciers results in the transfer of significant quantities of debris.
3. Deposition: Moraines, drumlins, and outwash plains are all examples of landforms that are formed as a result of glaciers melting and depositing sediments.

2. Periglacial Processes

A. Frost Action: The process of weathering and mass movement is caused by freeze-thaw cycles.

B. Permafrost Dynamics: The existence of permafrost has an effect on the stability of the soil as well as the geomorphic processes.

Biological Activity as a Geomorphic Agent

1. Vegetation

A. Root Action: Roots are able to penetrate rock and soil, causing them to break down and contribute to the process of weathering.

B. Organic Matter: The process of decomposition transfers organic material into the soil, which in turn affects the structure and stability of the soil.

2. Animals

A. Burrowing: Animals such as earthworms and rodents are responsible for the disturbance of soil and rock, which contributes to the creation of soil and facilitates erosion.

B. Bioturbation: The process of organisms mixing soil together leads to a change in the structure of the soil and encourages the cycling of nutrients.

3. Human Activities

A. Agriculture: Farming practices affect landscape modification and soil erosion.

B. Urbanization: Construction and development alter drainage patterns and natural landforms.

C. Mining and Deforestation: Significant changes in the landscape and increasing erosion are the results of extractive activity.

Chapter 32
Weathering and Erosion

Introduction

The term "weathering" refers to the process by which rocks, minerals, and soils are separated into smaller particles by a variety of processes that include chemical, biological, and physical processes.It plays a crucial role in shaping the Earth's surface and prepares materials for subsequent deposition and transportation by erosional processes. The several forms of weathering, the processes that cause them and the relationship between weathering and erosion are all discussed in depth in this chapter.

Types of Weathering

Weathering can be broadly categorized into three main types. Physical (mechanical), chemical and biological weatherings are the three primary forms of weathering that may be roughly classified based on their characteristics. When it comes to the disintegration of rocks and minerals, each kind includes a unique set of processes and mechanisms that contribute to the process.

A. Physical (Mechanical) Weathering:

The term "physical weathering" refers to the process by which rocks and minerals are broken down into smaller pieces without any chemical compositional changes occurring.

Alterations in temperature, the release of pressure, and the action of water, ice, and wind are the key ingredients that contribute to the process of physical weathering.

1. **Frost Wedging (Freeze-Thaw Weathering)**
   a. Process: When water seeps into fractures in rocks, it freezes. Since water expands when it freezes, it puts pressure on the rock, which ultimately causes it to fracture and finally break apart.
   b. Environment: Typically found in frigid areas that see numerous cycles of freezing and thawing.
2. **Thermal Expansion and Contraction**
   a. Process: When rocks are heated, they expand, and when they are not, they contract. Temperature fluctuations that occur repeatedly lead to stress, which ultimately results in cracking and fragmentation.
   b. Environment: Typically seen in arid settings that experience significant temperature shifts during the day.
3. **Exfoliation (Pressure Release)**
   a. Process: Erosion is responsible for the removal of overlaying rock, which results in a drop in pressure. This causes the underlying rock to expand and fracture in a direction that is parallel to the surface.
   b. Environment: Outcrops of granite and hilly locations are common places to find this.
4. **Abrasion**
   a. Process: The process of abrasion occurs when rocks and particles have a collision with one another, which causes the surface of the item to be worn away.
   b. Environment: It is found in settings that are influenced by wind, glaciers, and rivers.
5. **Salt Crystal Growth (Haloclasty)**
   a. Process: In the first step of the process, saltwater seeps into the pores of the rock and evaporates, leaving behind salt crystals that develop and apply pressure, which ultimately causes the rock to fracture.
   b. Environment: Common in desert and coastal regions of the world.

B. **Chemical Weathering:** Chemical weathering is the process which involves the alteration of the chemical composition of rocks and minerals through chemical reactions. It results in the formation of new minerals that are soluble ions and may easily be transported by water.

1. **Hydrolysis**
   a. Process: The process of hydrolysis involves the reaction of water with minerals, which results in the formation of new minerals and soluble ions. Clay minerals are formed, for instance, when the feldspar minerals found in granite engage in a reaction with water.
   b. Environment: It is typical in humid settings where there is an abundance of water.

2. **Oxidation**

a. Process: Oxidation is the process by which minerals react with oxygen, which generally results in the formation of oxides. Rust is an example of iron oxides that are formed when iron-containing materials undergo oxidation.

b. Environment: Frequent in regions that are exposed to the atmosphere.

3. **Carbonation**

a. Process: During the carbonation process, carbon dioxide combines with water to produce carbonic acid. This acid then reacts with minerals such as calcite in limestone, which ultimately results in the dissolution of the substance.

b. Environment: Found in karst settings that include a significant amount of limestone.

4. **Dissolution**

a. Process: The process of dissolution involves the dissolution of minerals in water, especially those minerals that are highly soluble, such as gypsum and halite (known as rock salt).

b. Environment: Typical in areas that experience high levels of precipitation and groundwater flow

5. **Hydration**

a. Process: During the process of hydration, minerals take in water, which causes them to soften and expand. Anhydrite, for instance, is capable of absorbing water and forming gypsum.

b. Environment: This phenomenon is seen in any climate where water is available.

C. **Biological Weathering:** Biological weathering involves the breakdown of minerals and rocks through the actions of living organisms. Plants, animals, and microbes may all play a role in the process, which can include both physical and chemical phenomena.

1. **Root Wedging**

a. Process: Plant roots grow into crevices in rocks and apply pressure as they expand, forcing the rocks to break apart. This process is known as root wedging.

b. Environment: It is common in all areas that are vegetated.

2. **Organic Acids**

a. Process: The breakdown of organic matter results in the production of acids, which are responsible for the chemical weathering of rocks and minerals.

b. Environment: It is typical in soils and regions that have a high level of biological activity.

3. **Microbial Activity**

a. Process: Bacteria and fungus are examples of microorganisms that create acids and other compounds that contribute to the process of chemical weathering.

b. Environment: It is often found in caves, rock surfaces, and soils.

4. **Animal Activity**

a. Process: Burrowing animals are responsible for the breakdown and decomposition of rocks and soils, which contributes to the process of physical weathering.

b. Environment: Common in regions where there are active populations of animals.

Erosional Processes Linked to Weathering

There are a number of natural forces, including water, wind, ice, and gravity, that are responsible for the process of erosion. Erosion is the process by which weathered rock and soil are shifted and removed from their original place. There is a close connection between weathering and erosion, with weathering being responsible for the breakdown of materials, which are subsequently moved by erosional processes.

A. **Water Erosion**

1. **Fluvial Erosion**

a. Process: Rivers and streams are responsible for the erosion of their banks and beds, which results in the transportation of materials farther downstream.

b. Features: V-shaped valleys, oxbow lakes, river meanders, alluvial fans, and deltas are the common features.

2. **Rain Splash Erosion**

a. Process: The impact of rainfall on bare soil causes particles to become dislodged, which are subsequently carried away by runoff from the environment.

b. Environment: It is typical in areas that experience heavy rainfall situations.

3. **Sheet Erosion**

a. Process: The process of sheet erosion involves the removal of thin layers of soil by surface runoff, which often occurs on slopes that are not steep (typically on gentle slopes).

b. Environment: This is typical in agricultural regions that do not implement adequate soil protection measures.

4. **Rill and Gully Erosion**

a. Process: The process of rill and gully erosion occurs when a concentrated flow of water begins to form tiny channels known as rills, which have the potential to grow into bigger gullies.

b. Environment: Typically found in regions that have soils that are prone to erosion and have minimal plant cover (insufficient vegetation cover).

B. **Wind Erosion**

1. **Deflation**

a. Process: Wind is responsible for the removal of loose particles from the surface, which ultimately results in the creation of desert pavements and deflation hollows.

b. Environment: Common in places that are either dry or semi-arid.

2. **Abrasion**

a. Process: Landforms such as ventifacts and Yardangs are formed as a result of the erosion of rock surfaces caused by wind-blown sand particles.

b. Environment: Deserts and coastal areas with loose sand are common places to see this.

3. **Dune Formation**

a. Process: Sand is deposited by wind in mounds or ridges, which results in the formation of dunes of a variety of morphologies, including barchan, transverse, and longitudinal dunes.

b. Environment: Deserts, coastal regions, and sandy habitats are all common places to find it.

C. Ice Erosion

1. **Glacial Erosion**

a. Process: Glaciers are responsible for the erosion of the terrain via plucking and abrasion, which results in considerable volumes of material being transported (debris).

b. Features: Fjords, U-shaped valleys, arêtes, cirques, and glacial striations are types of landscape features.

2. **Glacial Deposition**

a. Process: The melting of glaciers results in the deposition of sediments, which result in the formation of landforms such as moraines, drumlins, and eskers.

b. Environment: Common in mountainous and polar regions.

D. Gravity-Driven Erosion / Mass Wasting (Mass Movement)

a. Process: This is a movement of rock and soil that occurs downslope and is caused by the attraction of gravity.

b. Types:

- Falls: Materials that are falling freely through the air and moving at rapid speeds.
- Slides: Movement along a well-defined surface, such as rockslides and landslides.
- Flows: Movement as a viscous fluid, including debris flows, mudflows, and earth flows.
- Creep: The flow of dirt and rock downslope in a slow and steady manner.

c. Environment: Typical in areas that are steep and mountainous.

Implications and Applications of Weathering and Erosion Studies

A. Environmental Management

1. Soil Conservation: Implementing measures to prevent soil erosion in urban and agricultural areas.
2. Habitat Restoration: Understanding erosion and weathering processes to restore and protect natural habitats.

B. Natural Hazard Mitigation

1. Landslide Risk Assessment: Locating places that are prone to landslides and establishing ways to prevent them from occurring.
2. Flood Control: Designing flood defenses based on understanding fluvial erosion and sediment transport.

C. Engineering and Construction

1. Infrastructure Planning: When designing and constructing roads, bridges, and buildings, weathering and erosion processes should be taken into account.
2. Sustainable Development: Maintaining a balance between landform preservation and development.

D. Climate Change Studies

1. Impact Assessment: Studying how changing climate conditions affect weathering and erosion rates.
2. Adaptation Strategies: Developing strategies to adapt to climate-induced changes in landscape processes.

Difference between Weathering and Erosion

a. Mechanism of Action

1. Weathering: Involves the breakdown of minerals and rocks at their original location without movement. It can be chemical, biological or physical.
2. Erosion: Involves the removal and transport (movement) of weathered materials by natural agents such as wind, water, ice, and gravity.

b. Location and Mobility

1. Weathering: Occurs in situ (on-site), where the mineral or rock is located.
2. Erosion: Involves transporting, or movement of materials from one location to another.

c. Results and Effects

1. Weathering: Produces smaller particles, regolith, and soil; prepares materials for erosion.
2. Erosion: The process of transporting and depositing weathered materials, which results in the formation of new landforms and the alteration of landscapes.

d. Timescales

1. Weathering: The process is often slow and gradual, and it takes place over extended periods of time.
2. Erosion: Depending on the agent and the environment, changes may take place either quickly (for example, during floods) or gradually over a period of time.

Interactions between Weathering and Erosion

A. Complementary Processes

1. Weathering as a Precursor: Rocks are broken down into tiny bits by the process of weathering, which ultimately makes them easier for erosion to carry.
2. Feedback Mechanisms: Weathered material is removed by erosion, exposing new rock surfaces to more weathering.

B. Landscape Evolution

1. Dynamic Equilibrium: Landscapes achieve a balance between weathering, erosion, and deposition, continually reshaping throughout geological time spans.
2. Role of Climate and Vegetation: Climate and vegetation have a considerable impact on the rates of weathering and erosion, as well as the kinds of erosion that occur, which in turn shapes a variety of landscapes.

Case Studies Illustrating Weathering and Erosion

A. The Grand Canyon

1. Weathering Processes: Physical weathering, which is caused by variations in temperature, and chemical weathering, which is caused by precipitation and groundwater.
2. Erosional Processes: There are three types of erosion: mass wasting, rain splash erosion, and fluvial erosion caused by the Colorado River.

B. The Sahara Desert

1. Weathering Processes: These processes include thermal expansion and contraction, the formation of salt crystals, and biological weathering caused by the presence of restricted vegetation.
2. Erosional Processes: Abrasion, Wind deflation, and dune formation.

C. The Alps

1. Weathering Processes: Oxidation, Freeze-thaw weathering, and hydrolysis.
2. Erosional Processes: Mass wasting, fluvial erosion, and glacial erosion.

Implications and Applications of Weathering and Erosion Studies

A. Environmental Management

1. Soil Conservation: Implementing measures to prevent soil erosion in urban and agricultural areas.
2. Habitat Restoration: Understanding weathering and erosion processes to protect and restore natural habitats.

B. Natural Hazard Mitigation

1. Landslide Risk Assessment: The process of identifying regions that are prone to landslides and establishing ways to mitigate these.
2. Flood Control: Designing flood defenses based on understanding sediment transport and fluvial erosion.

C. Engineering and Construction

1. Infrastructure Planning: Considering weathering and erosion processes in the construction and design of bridges, buildings, and roads.
2. Sustainable Development: Striking a balance between the preservation of natural landforms and the development of land.

D. Climate Change Studies

1. Impact Assessment: The investigation of how the weathering and erosion rates are affected by the changing climatic conditions.
2. Adaptation Strategies: Creating plans to adjust to changes in landscape processes brought on by climate change.

Erosion and weathering are two different processes that alter the surface of the Earth, although they are deeply intertwined. Weathering is the process by which rocks and minerals are broken down in their natural environment, whereas erosion is the process by which these materials are transported to new areas. For environmental management, natural hazard mitigation, and sustainable development, it is essential to have a solid understanding of the interrelationships and distinctions between these systems. As climate change continues to impact the planet, studying weathering and erosion will be essential for predicting and adapting to future landscape changes.

Chapter 33
Geological Work of Wind and Their Landforms

Introduction to Aeolian Processes

The term "aeolian processes" refers to the volcanic and geological processes that are caused by the wind. The significance of these processes lies in the fact that they play a key role in the formation of dry and semi-arid regions, coastal areas, and other landscapes that have a limited amount of plant and accessible fine sediments. As a geomorphic agent, wind has the ability to erode, transfer, and deposit materials, resulting in the formation of a wide variety of different landforms. The mechanics of wind action, the several kinds of aeolian processes, and the landforms that are produced by these processes are discussed in this chapter.

Mechanisms of Wind Action

A. **Wind Erosion:** The removal and movement of soil and rock particles from one area to another is the process that is referred to as wind erosion. The efficiency of wind erosion is dependent upon the velocity of the wind, the size of the sediment, and the conditions of the surface.

1. **Deflation**: In the process of deflation, fine-grained particles that are free are carried away by the wind, leaving behind a surface material that is composed of coarse

particles or bedrock. Deflation hollows and desert pavements are produced as a result of it.

2. **Abrasion**: The process of wearing down surfaces by particles that are carried forward by the wind, similar to sandblasting. This results in the creation of yardings and ventifacts.
3. **Attrition:** The process of grinding is known as attrition, and during transit, particles that are created by the wind often clash with one another. An additional grinding of the particles is brought about as a result of such reciprocal impact.

B. **Wind Transportation:** Wind transports sediments through various mechanisms depending on wind speed and particle size.

1. **Suspension**: Fine particles (such as slit and clay) are lifted and carried over long distances in the air. It can lead to the formation of loess deposits and dust storms.
2. **Saltation**: Sand-sized particles are quickly raised into the air and then bounce along the ground. In deserts, saltation is the primary means by which sand is transported.
3. **Surface Creep**: Larger particles (such as gravel and sand) are pushed or rolled along the ground by impact from salting practices or by impact of. It is a factor that contributes to the slow movement of sand dunes.

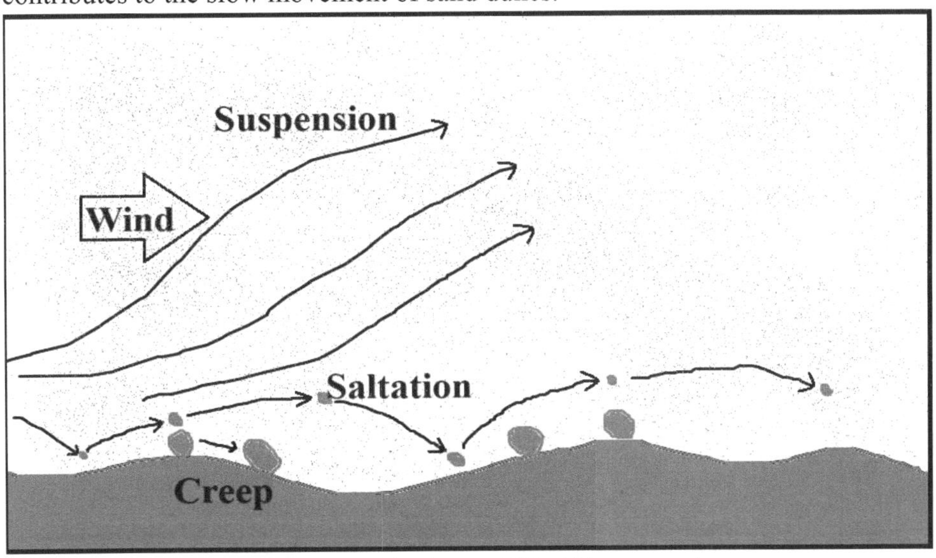

Figure 33.1: Mechanisms of Wind Transportation

C. **Wind Deposition:** As the wind's carrying capacity decreases, it dumps the sediments that it has carried, resulting in the formation of a variety of landforms.

1. **Sand Dunes**: Occur when wind-blown sand accumulates in places where the wind speed or velocity decreases. Types: Includes transverse, longitudinal, star dunes, parabolic, and barchans.

2. **Loess Deposits**: Silt and clay-sized particles that are wind-deposited and often found downstream of glacier and desert areas. Forms fertile soils and can create extensive, thick deposits.

Aeolian Landforms

A. Erosional Landforms

1. **Deflation Hollows**: Deflation hollows are low-lying depressions that form when wind erosion removes fine particles such as silt, clay and sand from the surface, leaving behind larger and heavier particles. The most prevalent conditions for this process, called deflation, are arid and semi-arid ones where loose sediments are prone to erosion. The smaller particles are blown away by the wind, often across wide regions, resulting in depressions. Examples: The Qattara Depression in Egypt.

Figure 33.2: Deflation Hollows erosional landforms formed by wind

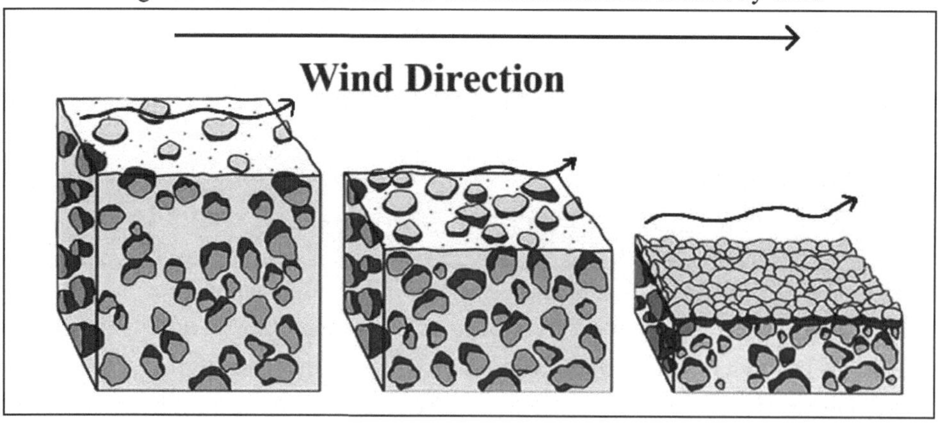

Figure 33.3: Desert Pavements erosional landforms formed by wind

2. **Desert Pavements**: A desert pavement, also called reg (in the Western Sahara), serir (eastern Sahara), gibber (in Australia), or saï (central Asia), is a layer

of rock pieces that are densely packed and interlocked. These fragments are angular or rounded in shape and are left behind after finer particles are removed. It is the effect of deflation and the presence of minimal vegetation.

3. **Ventifacts**: Rocks that have been etched, polished or shaped by wind-driven sand. Often have sharp edges and flat facets.

a. **Einkanters:** Ventifacts with one smooth surface.

b. **Zweikanter:** Ventifacts with two smooth surfaces.

c. **Dreikanter:** Ventifacts with three smooth surfaces.

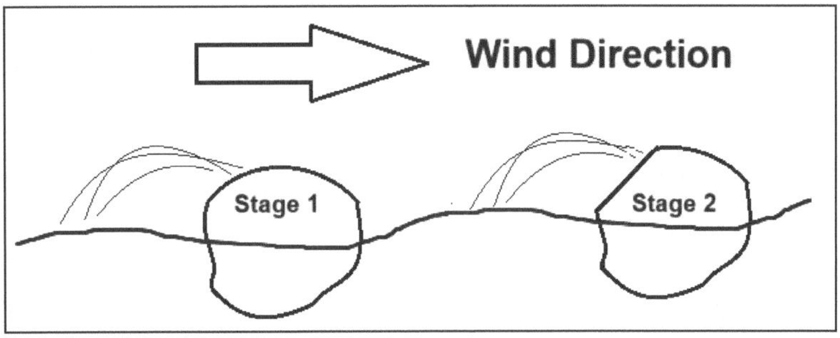

Figure 33.4: Ventifact erosional landforms formed by wind

4. **Yardangs**: Yardangs are wind-eroded, streamlined ridges that are usually parallel to the direction of the predominant wind. They are created in areas that are characterized by strong winds that are persistent and soft rock or silt that is prone to erosion. These places are favourable for the formation of rocks. Yardangs are formed by a combination of two processes: abrasion, which is the erosion that occurs when particles propelled by the wind strike the surface, and deflation, which is the removal of loose particles by wind. The ridges are defined by their typical stretched forms, which are created as a result of various activities throughout time.Example: Yardangs in the Lut Desert, Iran.

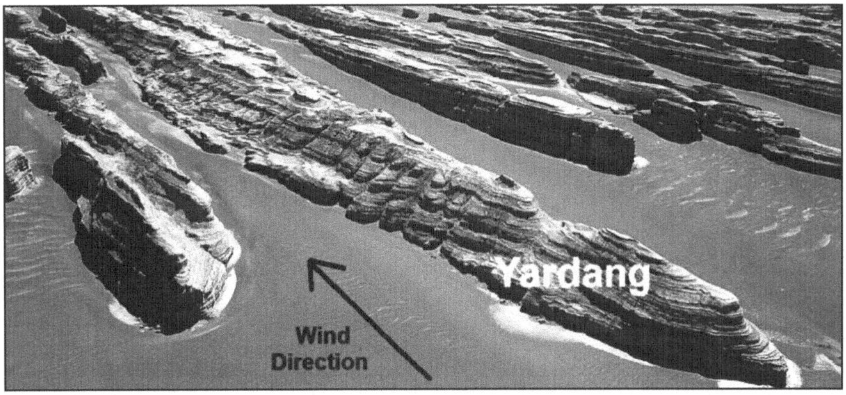

Figure 33.5: Yardangs erosional landforms formed by wind

5. **Hamada:** When the loose particles are carried away by deflation, only the hard mantle—known as "Hamada"—remains.

Figure 33.6: Hamada erosional landforms formed by wind

6. **Pedestal rock:** Pedestal rock, which is often referred to as mushroom rock, is a kind of geological formation that is distinguished by a huge rock cap that is broader in diameter and sitting on top of a narrower and more slender rock column or pedestal. These formations are mostly the result of wind abrasion, which occurs when particles carried by the wind, such as sand, erode the softer layers of rock at the base at a faster rate than the tougher layers at the top, resulting in the rock having the appearance of a mushroom.

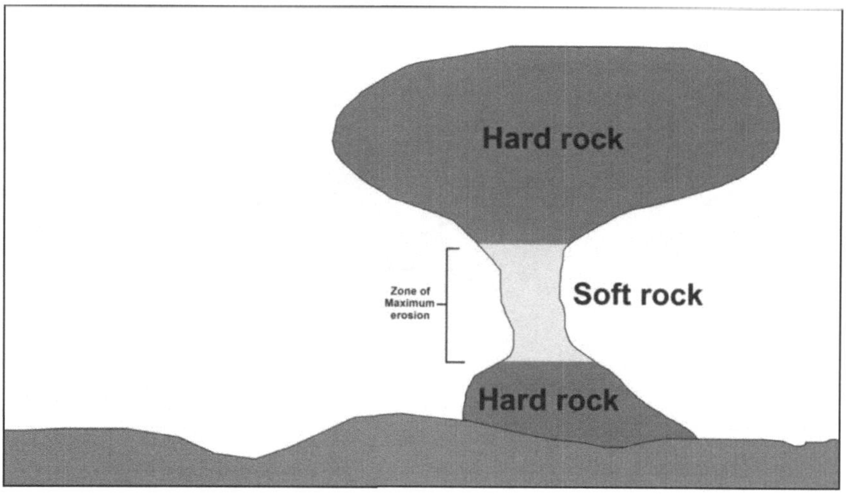

Figure 33.7: Pedestal rock erosional landforms formed by wind

7. **Honey-comb structure:** Honeycomb weathering, also known as honeycombs and honeycombed sandstone, is a type of cavernous weathering and a subcategory of tafoni. It is characterized by the presence of cavities that are regularly arranged, closely adjacent to one another, and frequently patterned. These cavities are formed in weathered bedrock and are less than 2 centimeters (0.79 inches) in size. Honeycomb weathering is named after the honeycombs that are formed within the bedrock. Honeycombs have also been referred to as alveoli, lacework, stonelace, fretting, and micro tafoni weathering.

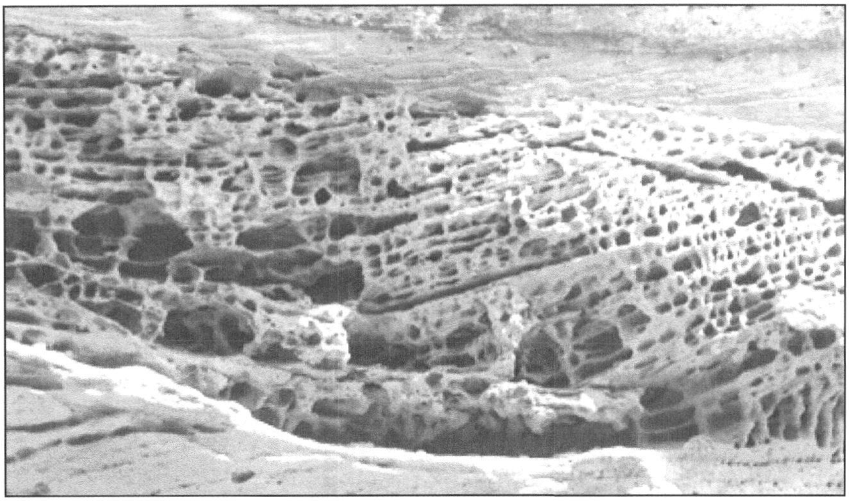

Figure 33.8: Honey comb rock erosional landforms formed by wind

8. **Mushroom table / Zeugen:** The structure is a tabular mass of more resistant rock that is supported by pillars of softer material that have been undercut. They often extend in the direction of the predominant wind.

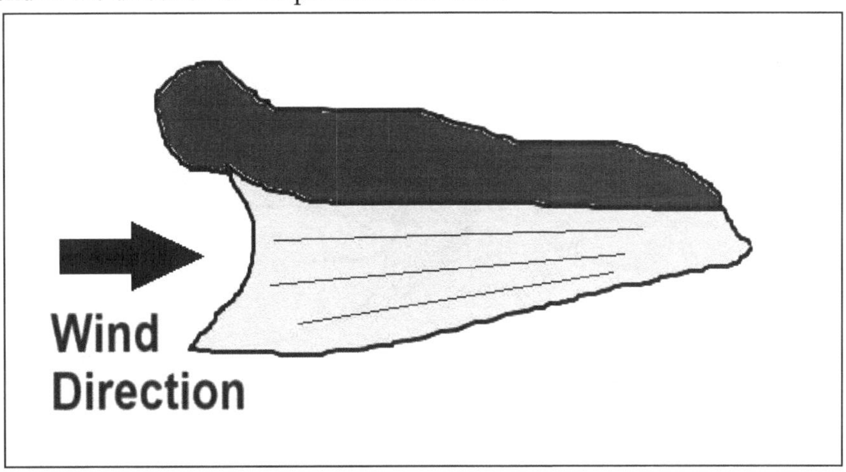

Figure 33.9: Zeugen / Mushroom Table erosional landforms formed by wind.

9. **Blow outs**: A blowout is a sandy depression that occurs in an ecosystem that is composed of sand dunes and is generated by the removal of sediments by wind. Blowouts are characterized by the formation of patches of bare sand atop vegetative dunes that have been stabilized by wind erosion. These blowouts are often seen in coastal areas and adjacent desert regions.

Figure 33.10: Blowouts erosional landforms formed by wind

10. **Millet seed sands:** Sand that is composed of grains that are smooth and spherical, about the size of a millet seed, and is often suggestive of extended aeolian activity that was caused by the process of attrition.

1. **Depositional Landforms**

1. **Sand Dunes**

a. **Barchan Dunes:** Barchan Dunes are a kind of dunes that have a crescent shape with horns that point downwind. These dunes are the result of a limited amount of sand and a wind that blows in only one direction.

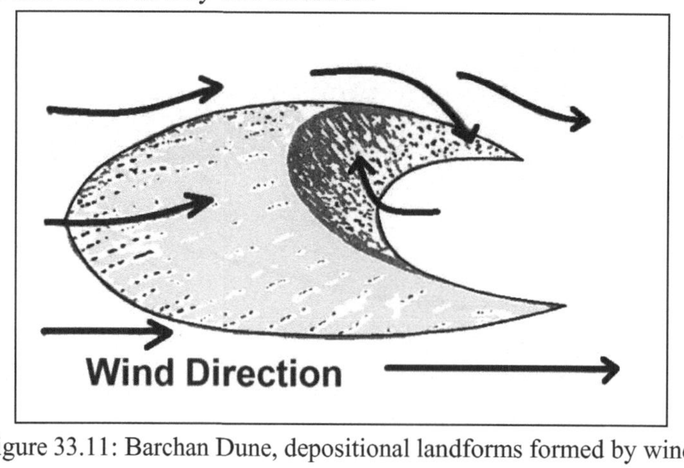

Figure 33.11: Barchan Dune, depositional landforms formed by wind.

b. **Seif**: It resembles a barchan, with the exception that one wing is absent due to periodic changes in wind direction.

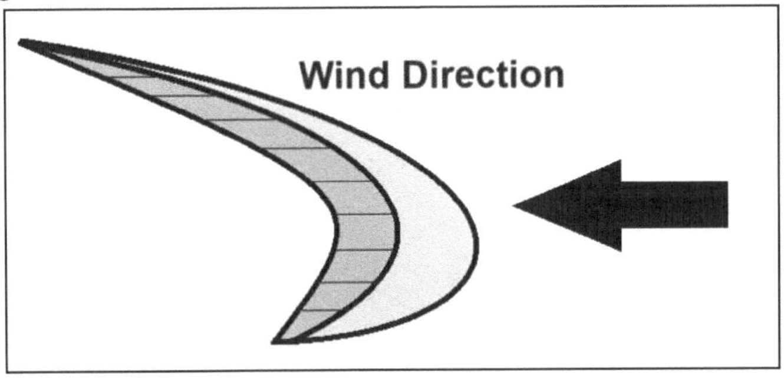

Figure 33.12: Seif Dune, depositional landforms formed by wind

c. **Transverse Dunes:** Found in regions with a steady wind direction and an abundance of sand, these large, linear dunes are aligned perpendicular to the prevailing wind.

d. **Longitudinal Dunes:** In regions with a modest supply of sand and bidirectional winds, longitudinal dunes—long, thin ridges parallel to the prevailing wind—form.

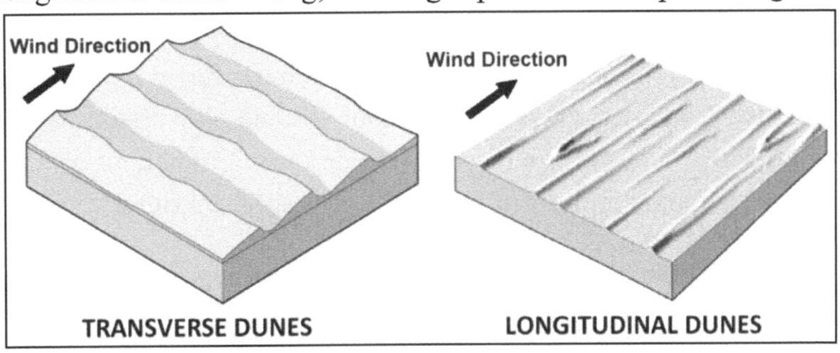

Figure 32.13: Transverse Dunes and Longitudinal Dune, depositional landforms formed by wind.

e. **Parabolic Dunes:** U-shaped dunes with tips pointing upwind; stabilized by vegetation and form in areas with some moisture.

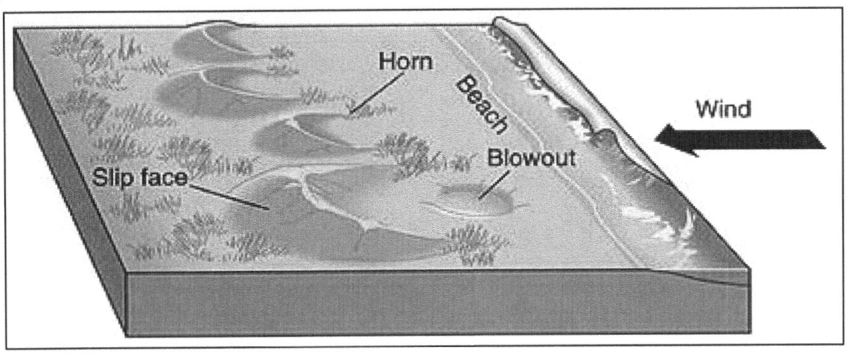

Figure 33.14: Parabolic Dune depositional landforms formed by wind (Reference: *Wikipedia, The Free Encyclopedia*)

f. **Star Dunes:** Star dunes are classified as radially symmetrical dunes that have several arms and are formed in regions that experience varying wind directions.

Figure 33.15: Star Dune a depositional landforms formed by wind

2. **Loess Deposits**: It consists of vast blankets of silt and clay that have been deposited by the wind. There is a prevalence of loess deposits throughout the central United States, China, and some regions of Europe. Loess soils are often very productive and play an essential role in agricultural production.

Figure 33.16: Loess Deposits

3. **Sand Sheets**: Sand sheet are large, flat stretches of sand that are generally seen on desert edges. They are deposited by wind. This phenomenon takes place when the wind speed drops across a large region, which results in the settling of sand.

Figure 33.17: Sand Sheets

4. **Ripples**: Ripples are a kind of wave-like characteristic that may be seen on the surface of sand or loose dirt. The flow of wind over loose sand causes grains to hop or roll and gather in wave-like patterns, which results in the formation of ripples. The speed of the wind and the size of the sand grains both have an effect on the size of the ripples and the distance between them.
5. **Draa:** Draa are large-scale dune complexes that may measure tens of kilometers in length and are far bigger than individual sand dunes. Draa are formed in desert environments that have a high concentration of sand and wind patterns that remain continuous over extended periods of time. This allows for the construction of vast dune systems.
6. **Erg (Sand Sea):** An erg is a large region that is covered with sand dunes and is often referred to as a sand sea. Sand is deposited across huge areas in dry locations, resulting in the formation of ergs. Within these ergs, dunes of a variety of kinds are often seen. They are the epitome of aeolian deposition in arid settings, and they represent the peak of the phenomenon.

2. **Coastal Aeolian Landforms**

1. **Coastal Dunes**: Coastal dunes are formed when sand that has been removed from beaches and blown inland accumulates. It is often stabilized by flora, such as grasses, which contribute to the sand's ability to bond together.
2. **Sand Spits and Bars**: A combination of wind activity and longshore drift may result in the formation of thin ridges of sand that extend into the sea. Examples: Spits like Spurn Point in the UK.

3. **Fore dune:** Ridge-like deposits of wind-blown sand that are created along the shore of lakes or the ocean are known as fore dune formations.
4. **Beach Ridges**: The thin, elongated ridges of sand or gravel that run parallel to the shoreline are known as beach ridges. Their formation is the result of the interaction between the wind and the waves. These ridges are further shaped and elevated by wind after waves have deposited sand on them, which often results in an increase in both their height and breadth.
5. **Whale back dune-**It is a massive, flat-topped longitudinal dune that may be home to barchans or seifs.

The area between the dunes is referred to as **"Gassis,"** while the water that supports plants in arid regions and is accessible in shallow wells is known as **"Oasis."**

Factors Influencing Aeolian Processes

A. Wind Speed and Direction

a. Velocity: The potential for erosion and transportation is increased significantly when wind speeds are higher.
b. Direction: Wind directions that are consistent result in landforms that are more prominent and ordered.

B. Sediment Availability and Size

a. Supply: An important factor that determines the level of aeolian activity is the quantity of loose material that is readily accessible.
b. Particle Size: Coarser particles need greater wind speeds in order to be raised and moved, while finer particles are more readily lifted and transported.

C. Vegetation Cover

a. Stabilization: Sediments may be anchored by vegetation, which assists in the creation of specific landforms such as parabolic dunes and reduces the amount of erosion that occurs.
b. Loss of Vegetation: Increases vulnerability to erosion caused by wind.

D. Climate and Moisture

a. Aridity: Due to the absence of moisture that binds particles together, dry conditions are more conducive to wind erosion and movement.
b. Moisture: Even small amounts of moisture can significantly reduce wind erosion by increasing soil cohesion.

Human Impact on Aeolian Processes

A. Land Use Changes

a. Agriculture: Overgrazing and plowing may reduce plant cover, making an area more vulnerable to wind erosion.
b. Deforestation: Forest clearing exposes the soil to wind.

B. Urbanization and Infrastructure

a. Construction: Wind erosion may be increased by construction-related disturbance of the soil and sand.

b. Roads and Buildings: May change wind patterns and have an impact on the deposition and transportation of sediments.

C. Climate Change

a. Droughts: Wind erosion may worsen if droughts occur more often and with greater intensity.

b. Desertification: Human activities and climate change have led to the expansion of desert regions, which has resulted in an increase in the area that is influenced by aeolian processes.

Case Studies of Aeolian Processes and Landforms

A. The Sahara Desert

a. Deflation and Abrasion: Extensive areas of deflation hollows and desert pavement.

b. Sand Dunes: Large expanses of sand dunes that include a variety of sand dunes, such as barchan and stars.

B. The Dust Bowl (1930s, USA)

a. Human Impact: High levels of wind erosion were caused by over cultivation and drought.

b. Effects: The depletion of topsoil and the occurrence of massive dust storms have caused major harm to both agriculture and the economy.

C. The Loess Plateau (China)

a. Loess Deposits: The term "loess deposits" refers to thick layers of wind-blown silt that create soils that are fruitful yet prone to erosion.

b. Agricultural Impact: Extensive terracing to combat soil erosion and maintain agricultural productivity.

Mitigation and Management of Wind Erosion

A. Vegetation Cover

a. Planting: Establishing shrubs, trees or grasslands to reduce wind speed and stabilize soil at the surface.

b. Windbreaks: Planting rows of shrubs, bushes or trees to act as barriers against wind.

B. Soil Management

a. Mulching: Applying a layer of organic material to the surface of the soil in order to preserve it is known as mulching.

b. Conservation Tillage: Minimizing the amount of soil disturbance in order to preserve the structure of the soil and cut down on erosion.

C. Land Use Planning

a. Avoiding Overgrazing: Managing livestock to prevent soil exposure and vegetation loss.
b. Sustainable Agriculture: The implementation of techniques that aim to preserve the health of the soil and prevent the danger of erosion.

D. Technological Solutions

a. Sand Fences: Structures intended to stabilize and hold onto sand in regions where dune migration is likely to occur.
b. Erosion Control Mats: Natural or synthetic mats that protect the soil surface from wind action.

The geological work that wind does is an important factor in the formation of a variety of landscapes, especially in locations that are dry or semi-arid. Wind is responsible for the formation of a wide range of distinctive landforms, such as dunes, loess deposits, and ventifacts, via different processes like as erosion, movement, and deposition. When it comes to controlling and reducing wind erosion, having a solid understanding of these processes and the consequences they have on the environment is very necessary, particularly in light of the fact that human activities and climate change are causing it. Through the study of aeolian processes, we are able to increase our understanding of the dynamic character of the surface of the Earth as well as the continual interplay between natural forces and the acts of humans.

Chapter 34
Geological Work of Rivers and Their Landforms

Introduction to Fluvial Processes

Rivers are inherently dynamic systems that play a significant part in the process of altering the surface of the Earth. Rivers are responsible for shaping a variety of landforms and for having an impact on landscapes via the processes of erosion, movement, and deposition. In this chapter, we investigate the geological work that rivers do, the mechanisms that are responsible for fluvial processes, and the landforms that are produced as a result of these activities.

Mechanisms of Fluvial Processes

A. **River Erosion:** Rivers use a variety of processes to erode their channels and the landscapes that surround them. These processes include hydraulic action, abrasion, solution, and attrition.

1. **Hydraulic Action**: The force of running water removes loose material from the river bed and banks, which results in erosion of the river bed. Hydraulic action is a phenomenon that often occurs in streams with high velocity and during flood occurrences.
2. **Abrasion**: The bed and the banks of the river are worn away as a result of the sediments and rocks that are carried by the river. As a result of its effects, the rocks become smoother, and the river channel becomes deeper.

3. **Solution (Corrosion)**: The soluble minerals that are found in the river bottom and banks are dissolved by water. It is often found in regions that have limestone and other minerals that are soluble.
4. **Attrition**: The particles that are transported by the river come into contact with one another and shatter into smaller bits. As they flow downstream, the size of the particles gradually decreases.

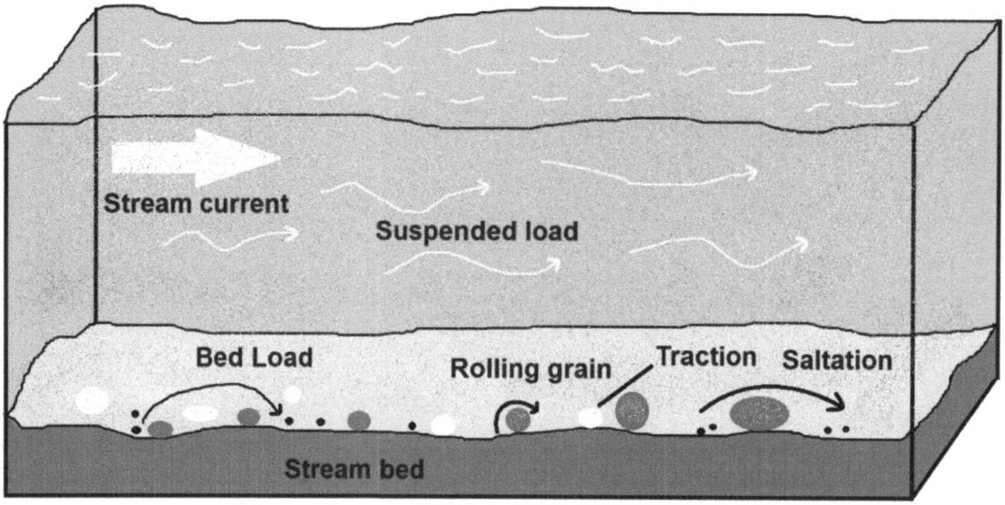

Figure 34.1: Transportation of sediments in river.

B. **River Transportation:** Sediments are transported by rivers via a variety of processes, which are affected by the velocity of the water and the particle size.

1. **Traction**: The river bed is followed by the rolling or dragging of large particles such as stones and pebbles. High-energy settings, such as mountain streams, are characterized by the presence of traction.
2. **Saltation**: The particles, which are about the size of sand, are raised and then bounce down the river bed. In settings with moderate to high levels of energy, saltation is a regular occurrence.
3. **Suspension**: In the water column, fine particles such as silt and clay are transported from one place to another. It is normal for suspension to occur in surroundings with low to moderate levels of energy.
4. **Solution**: In water, minerals that have been dissolved are transported. In all kinds of water bodies, solution may be found, although it is most prevalent in those places where water chemistry supports dissolving.

C. **River Deposition**

When a river reaches its carrying capacity, it continues to deposit sediments, which results in the formation of a variety of landforms.

1. **Decrease in Velocity**: When the water slows down, the heavier particles are the ones that are deposited first. The floodplains, deltas, and river mouths are all frequent places to find it.
2. **Change in Gradient**: If the gradient is lowered, then the amount of energy and sediment that is deposited will also decrease. It is typical in regions that are transitioning from highland to lowland environments.
3. **Obstructions**: Obstructions, whether they are natural or manufactured, are the source of localized deposition. Typically, it may be found hiding behind stones, foliage, and dams.

A theoretical model that attempts to explain the development of a river and its surrounding terrain over the course of geological time is called the **fluvial cycle of erosion.** This idea was first put up by William Morris Davis in the latter part of the 19th century, and it is sometimes referred to as the **"Davisian cycle of erosion."** Specifically, the concept proposes that a river passes through a series of phases, beginning with its youth and ending with its old age (youth, maturity and old age). Each of these stages is distinguished by distinct landscape characteristics and activities. Each of these steps is broken down as follows:

1. **Initial Stage:** During the first stage of river development, which includes the formation of the river itself, several processes that are distinctive of rivers include:Deepening of the channel as a result of bed-scouring, drilling potholes (via whirling current action),tributaries are quickly created, head ward erosion is maximum, Canyons, gorges, and waterfalls are formed.
2. **Youth Stage:** There are a number of processes that are indicative of the river's creation that take place during the Youth stage.River capture, also known as river piracy, is a phenomenon that occurs when one river reaches the path of another river. If the course of the second river is diverted due to the larger gradient of the earlier river, then this occurrence is referred to as river piracy. Elbow of capture refers to the location when the second river's course is diverted. The captured river is called the misfit-river.The wind-gap is the section of the channel that has been abandoned and through which no water flows. In this stage, there is noticeable formation of V-shaped valleys. The landscape is rugged with steep slopes. The river has a strong gradient, which might result in significant down cutting or erosion in a vertical direction. Valleys are typically in a V-shape. During the process of the river eroding through layers of refractory rock, waterfalls and rapids are often seen. It is dominated by vertical erosion. The river cuts into the landscape, deepening its channel.
3. **Mature Stage:** When rivers reach the mature stage of their creation, they go through a number of processes that are distinctive of rivers, such as: The gradient of the river begins to reduce, and it starts to meander over the landscape. As a result of lateral

erosion, also known as sideways erosion, valleys expand and take on a U-shaped form. The flow of the river starts to deposit silt, and floodplains begin to form. The topography is less rough and has a reduction in elevation (slope). The significance of lateral erosion increases in comparison to that of vertical erosion. The amount of sediment that is deposited rises, particularly during floods.

4. **Old Stage**: There are a number of processes that are indicative of the Old stage of river development, which includes: The River runs through a floodplain that is large and flat. The meanders grow more apparent and may contribute to the formation of oxbow lakes. The terrain is characterized by low topography, with narrow valleys that are flat in the middle. The course of the river is relatively slow, and it has a very low grade. It is dominated by deposition rather than erosion. The floodplain is progressively enlarged by the river's constant sedimentation.
5. **Rejuvenation:** Several distinctive mechanisms are involved in the river development process during the Rejuvenation stage: When there is a rise in the elevation of the land or a fall in the level of the sea, rejuvenation takes place. This causes the river to enhance the amount of erosive force it has. The river may initiate a new cycle of erosion, which will result in the formation of features such as river terraces and the cutting away of its floodplain. Revitalized vertical erosion process, which contributes to the formation of a more youthful landscape inside an older and more mature system.

Limitations of the Model:

a. The Davisian cycle of erosion is a model that is oversimplified and idealized. Numerous elements, including as temperature, tectonic activity, and differences in rock type, all have an impact on the landscapes that exist in the real world. These influences may not fit neatly into this cycle.

b. The cycle of erosion is often regarded by contemporary geomorphologists as one of the many viable frameworks for explaining the development of landscapes. This is due to the fact that landscapes do not always change in such a linear or predictable fashion.

Fluvial Landforms Created by Erosion

The passage of water in rivers and streams, which wears away soil and rock over time, is responsible for the formation of fluvial landforms, which are formed by erosion. Here are some key fluvial landforms that are result from erosion:

1. **V-Shaped Valleys**: Rivers cause the ground to erode vertically, resulting in valleys that are narrow and steeply sloped. It occurs often in highland regions that have a steep gradient and a high rate of water flow. Examples: Grand Canyon (USA), Tiger Leaping Gorge (China).

Figure 34.2: V-shaped valley, showcasing its steep sides and narrow floor created by fluvial erosion.

2. **Interlocking Spurs:** Ridges and valleys are created in alternating patterns as a result of river erosion and meandering around resistive rock. It is commonly found in the upper courses of rivers. Examples: Ingleton Waterfalls Trail (UK).

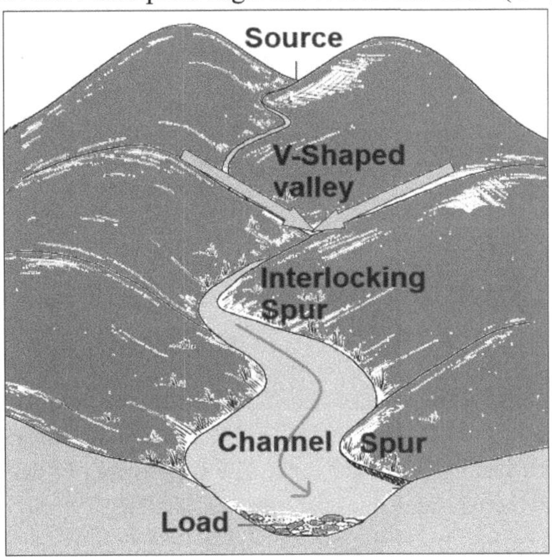

Figure 34.3: a river valley with interlocking spurs, highlighting the distinct zigzag pattern and the river winding through the narrow valley floor.

3. **Waterfalls, Rapids and Knick Points:** Knickpoints are points along rivers or streams that have an abrupt change in gradient, often a steep drop, and are frequently

indicated by waterfalls or rapids. Changes in rock resistance, tectonic activity, or fluctuations in the base level (for example, sea level or lake level) are the factors that may lead to the formation of knickpoints. Erodes the river bed by abrasion and undercuts the soft rock with the use of hydraulic action at the same time. Examples: Niagara Falls (USA/Canada), Angel Falls (Venezuela).

Figure 34.4: Image depicting a waterfall, rapids, and Knick Points formed where there is a sudden change in gradient and a band of resistant rock. The scene captures the powerful movement of water as it cascades over the falls and churns through the rapids.

4. **Gorges:** Created either by the gradual retreat of waterfalls or by the continuous erosion of steep surfaces. In most cases, it may be discovered in areas that are downstream of waterfalls or in locations that have resistant rock formations. Examples: Grand Canyon (USA), Iron Gates (Romania/Serbia).

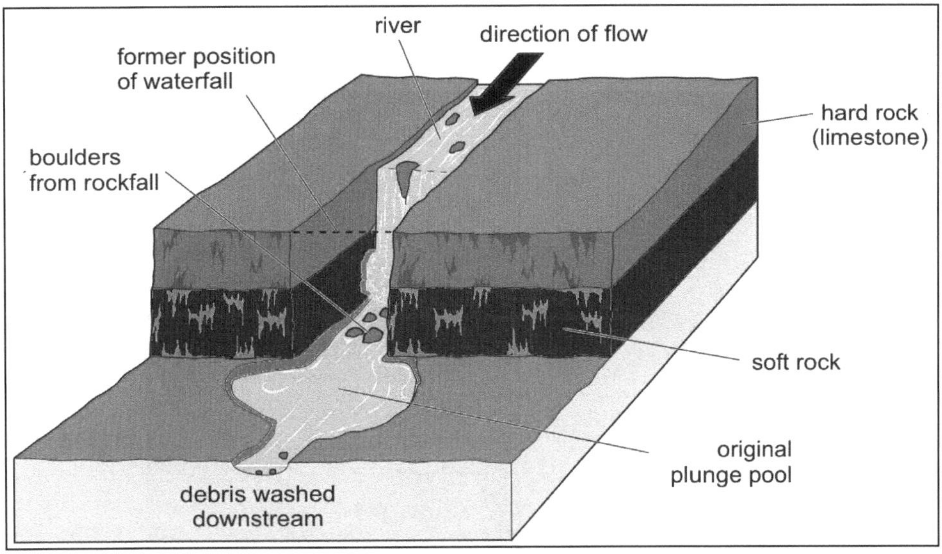

Figure 34.5: Image of a river gorge, showcasing its deep, narrow valley with steep rock walls and a winding river at the bottom

5. **Meanders:** Sinuous river routes are created as a result of lateral erosion and deposition in the middle and lower courses of the river. Hydraulic activity and abrasion lead to erosion on the outer bend of the structure. As a result of decreased velocity, sediment is deposited on the inner bend, which leads to sediment deposition. Examples: River Severn (UK), Mississippi River (USA). Par River (India)

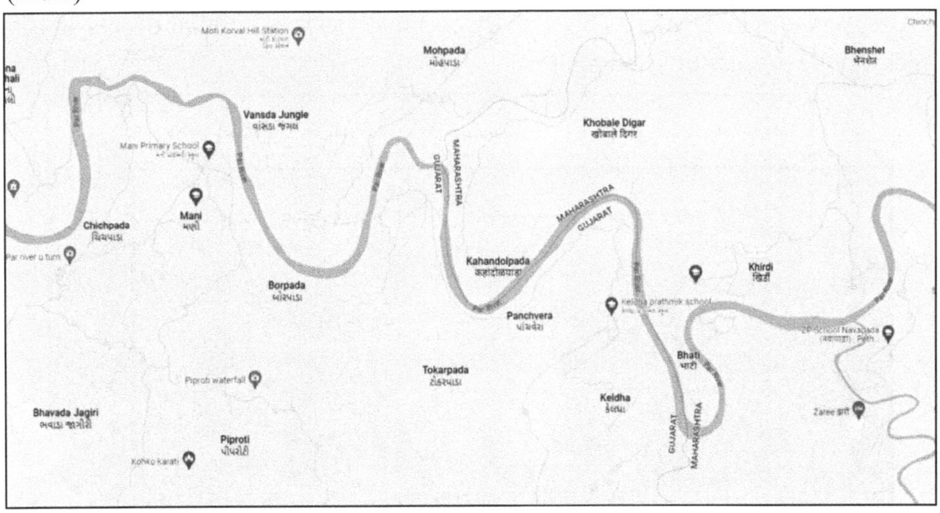

Figure 34.6: diagram showing Meanders in Par River near Maharashtra- Gujarat boarder

6. **River Cliffs:** A river cliff is a steep bank that is located on the outside of a river bend. This kind of bank causes the water to flow more quickly, which in turn causes

the bank to erode more quickly. Lateral erosion is the process by which these cliffs are formed. This process occurs when the river cuts into the outer bank of a meander.

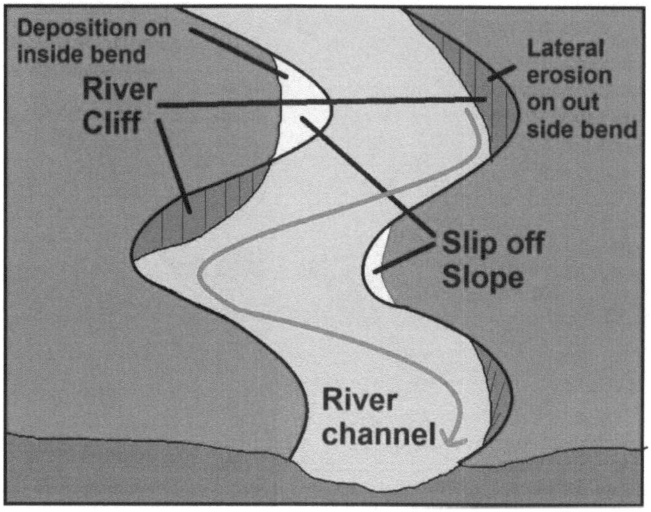

Figure 34.7: Diagram showing River Cliffs and Slip-off Slopes.

7. **Slip-off Slopes**: There are gentle slopes on the inside of a river bend that are known as slip-off slopes. These slopes are where sediment is deposited. Deposition is the result of the slower flow on the inner side of the curve, which was responsible for the formation of these slopes.
8. **Potholes:** Potholes are holes that are bored into the bedrock of a riverbed and have a cylindrical shape. It is the grinding action of pebbles and silt that swirls in waves that causes potholes to develop. This activity eventually erodes circular depressions in the rock via the process of grinding.

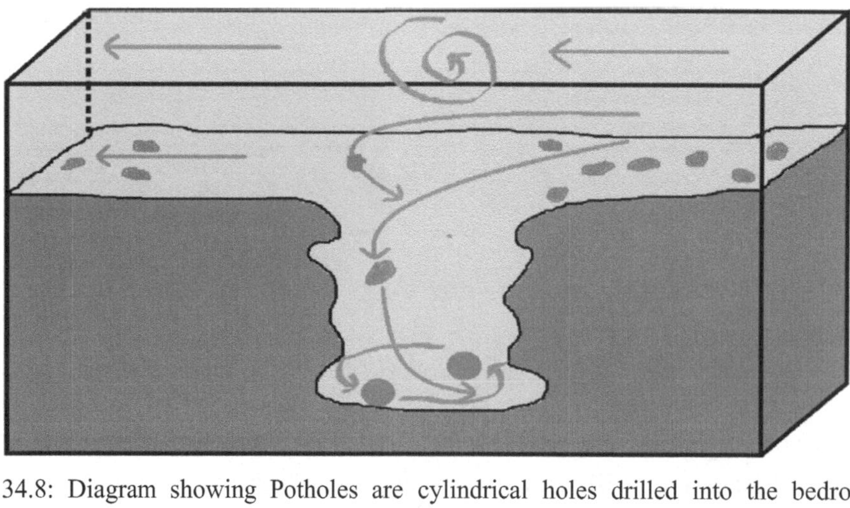

Figure 34.8: Diagram showing Potholes are cylindrical holes drilled into the bedrock of a riverbed.

9. **River Terraces**: Located above the current level of a river, river terraces are a kind of landform that resembles steps and represents floodplains that were formerly existent. As the river carves a deeper channel, it leaves behind sections of the previous floodplain, which results in the formation of terraces. Terraces are developed as a result of the river's down cutting operation.
10. **Strath**: A strath is a sort of river terrace, which is a landform that resembles a bench and is flat and develops along the banks of a river. Straths are formed when a river cuts laterally into the environment that surrounds it, resulting in the formation of a level surface above the river.As the river continues to erode downhill over the course of time, the strath will eventually be left raised above the level at which the river is now flowing.

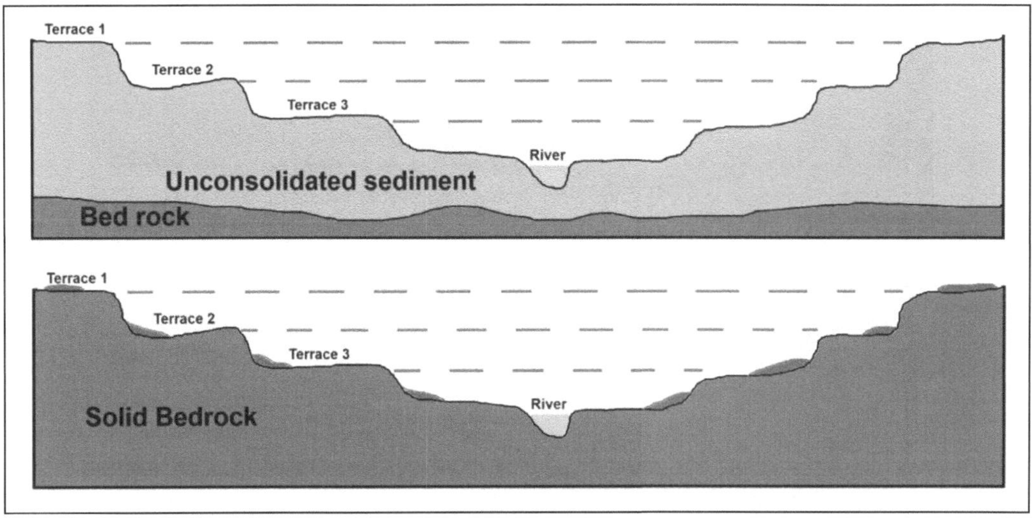

Figure 34.9: Diagram showing step-like landforms of River terraces

11. **Bedland**: Bedland refers to the eroded landforms or bedrock that expose bedrock.Develop on argillaceous rocks, which are a substance similar to clay, as a result of erosion caused by rushing water in the places that are passed by gullies.The term "badlands" refers to a kind of arid terrain that is characterized by the severe erosion of clay-rich soils and softer sedimentary rocks by wind and water. The steep slopes, the lack of vegetation, the absence of a large regolith, and the high drainage density are the defining characteristics of these areas.
12. **Escarpment**: An escarpment is a high slope or long cliff that divides two relatively flat regions of different heights. It is formed as a consequence of erosion or faulting and may run for a considerable distance. Erosion of sedimentary rock strata is a common process that results in the formation of escarpments.

Figure 34.10: Diagram showing bedrock or the eroded landforms

Figure 34.11: Diagram showing an escarpment is a steep slope or long cliff that results from erosion.

13. **Cuesta**: A hill or ridge that has a moderate slope on one side and a steep slope on the other is called a cuesta. This kind of hill or ridge is generated by the differential erosion of sedimentary rock strata that are inclined vertically. While the steep slope is located on the scarp side of the rock layers, the gentle slope is located on the dip side of the rock layers.

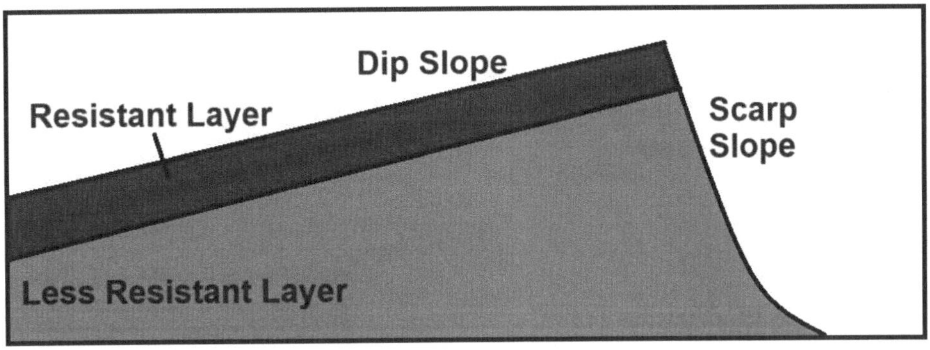

Figure 34.12: Diagram showing a cuesta hill or ridge with a gentle slope on one side and a steep slope

14. **Mesa**: A mesa is a peak or hill that is solitary, has a flat top, and steep sides. Mesas are often found in dry climates. Mesas are produced when sedimentary layers are eroded, leaving behind a layer of rock that is resistant to erosion on top, which shields the subterranean layers that are more fragile.
15. **Butte**: When compared to a mesa, a butte is on the smaller side. It is a hill that is completely isolated, with steep sides that are often vertical, and relatively small, flat summit. The erosion of mesas or other bigger landforms is the process that results in the formation of buttes.
16. **Hogback**: A hogback is a narrow ridge that has steep slopes on both sides. It is formed by the differential erosion of rock strata that dip sharply with respect to one another. When the rock strata are struck, the ridge is often aligned with the strike.

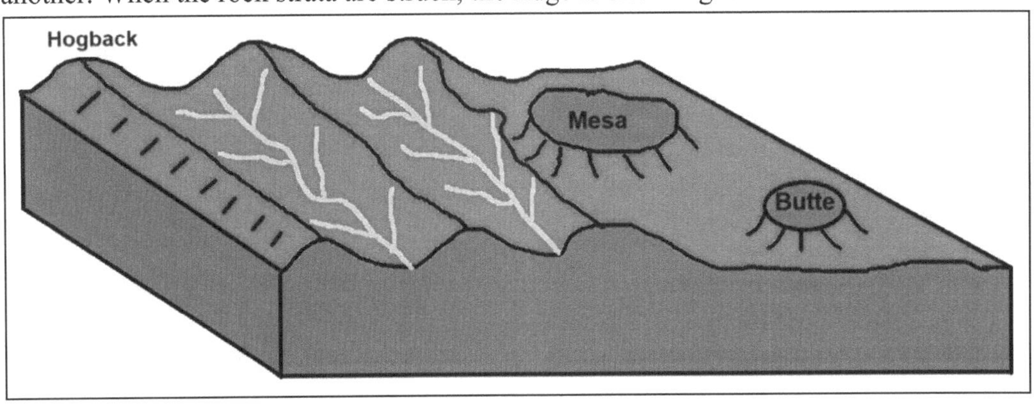

Figure 34.13: Diagram showing Mesa, Butte and Hogback.

17. **Peneplains**: A peneplains is a large, almost level land surface that represents the ultimate stage of fluvial erosion that occurs during periods of low tectonic activity. This stage occurs when the terrain is eroded down to a plain that is close to sea level.
18. **Monadnock**: A Monadnock, also known as an inselberg, is a peak or hill that is located in a remote area and rises suddenly from a surrounding plain that is either gently sloped or virtually flat. Monadnocks are often made up of more indestructible granite that is resistant to erosion, while the surrounding softer rock gets worn away.

Figure 34.14: Diagram showing Mesa, Butte and Hogback.

Fluvial Landforms Created by Deposition

1. **Floodplains:** Alluvium is deposited by periodic floods which results in the formation of fertile plains that are flat. The formation of floodplains occurs when a river exceeds its banks during times of high discharge. The river then stretches out over the floodplain, depositing silt as the water velocity lowers. The floodplain is built up over time by the accumulation of layers of silt, sand, and other materials that are deposited as a result of periodic floods. Agriculture relies heavily on the presence of fertile soils. Examples: Nile Delta (Egypt), Ganges Delta (India/Bangladesh).

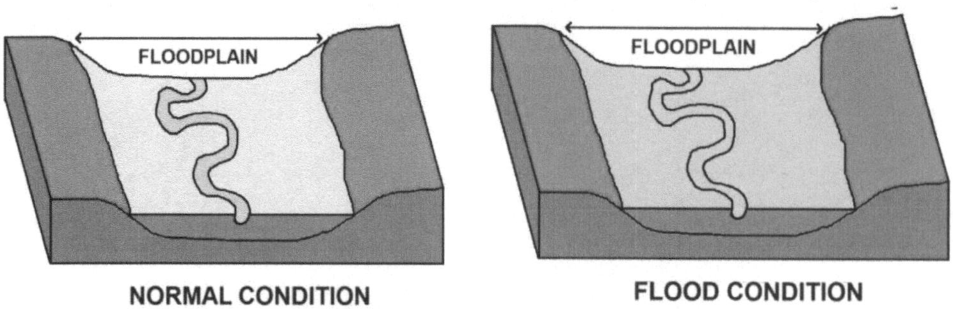

Figure 34.15: Diagram showing floodplains in normal and flood condition.

2. **Levees:** Located along the banks of a river channel, levees are natural embankments or high banks that are constructed. During floods, the river exceeds its banks, and as the water spreads out, its velocity diminishes. This causes the heaviest sediments to be deposited closest to the river channel, which results in the formation of levees. During succeeding floods, these sediments will eventually accumulate to create natural levees, which will assist in containing the river. Examples: Mississippi River (USA), Yellow River (China).

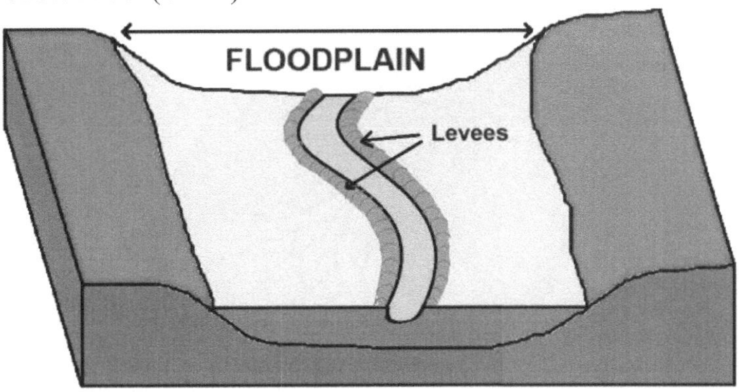

Figure 34.16: Levees along the sides of a river channel.

3. **Deltas**: A region of land that is triangular or fan-shaped is called a delta. It is a landform that is located at the mouth of a river, where the river empties into an ocean, sea, lake, or another river, depositing sediments that produce the delta. Deltas are produced when a river meets a body of water that is relatively still. This causes the river's velocity to drop significantly, which in turn causes sediments to settle out and collect. This delta is formed by the accumulation of sediments over time, which often results in the formation of several channels or distributaries.

Types of Deltas:

a. Arcuate Deltas: Fan-shaped, with branching distributaries (e.g., Nile Delta).
b. Bird's Foot Deltas: Protrude into the body of water (e.g., Mississippi Delta).
c. Cuspate Deltas: Pointed, with a few distributaries (e.g., Tiber Delta).

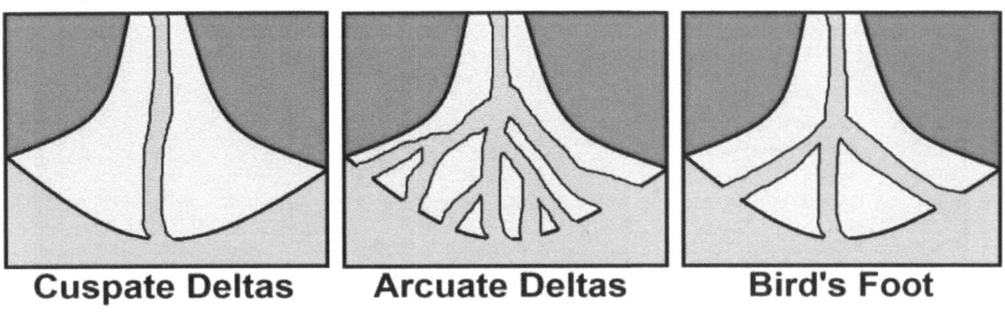

Figure 34.17: Diagram showing Cuspate, Arcuate and Bird's Foot Deltas

4. **Alluvial Fans:** Fan-shaped deposits of sediment are known as alluvial fans, and they are formed when a river or stream rushes out of a steep and narrow valley onto a plain that is flatter. As the water rushes out of the constricted valley and onto the plain, its velocity drops, which causes it to deposit the coarser materials (such as gravel and sand) closer to the valley entrance and the finer sediments farther away, resulting in the formation of a fan-shaped deposit. Examples: Death Valley (USA), Andes Mountains (South America).

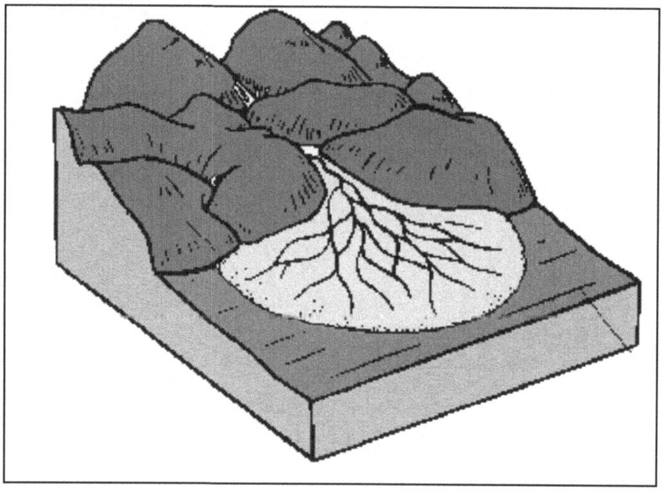

Figure 34.18: diagram showing alluvial fans: fan-shaped deposits of sediment.

5. **Point Bars:** Point bar deposits are deposits of sediment that are formed like crescents and may be found in the inner bends of river meanders. During the process of the river flowing around a bend, point bars are formed. This is because the water on the inside of the curve travels more slowly, which allows sediments to be deposited. This kind of deposition results in the formation of a point bar, which expands over time as more sediment is deposited.

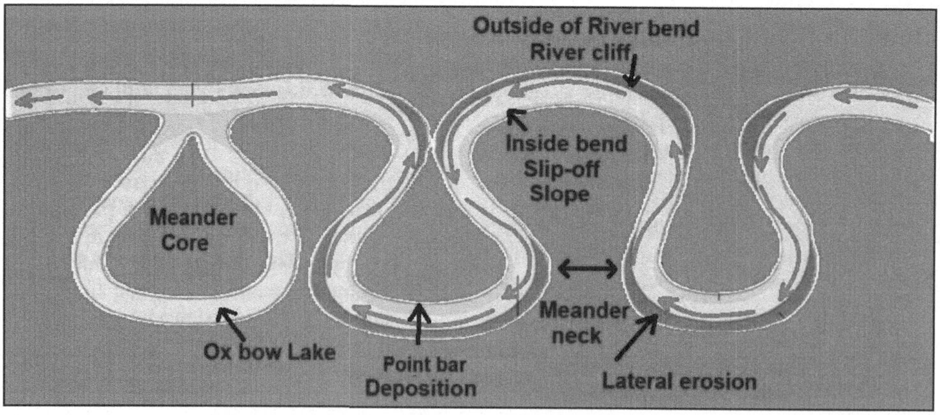

Figure 34.19: Diagram showing mature stage of river: ox-bow Lake, Point bar.

6. **Oxbow Lakes:** The formation of a U-shaped body of water known as an oxbow lake occurs when meanders are severed from the main channel, resulting in the formation of lanes that are crescent-shaped. As the meander of a river gets increasingly prominent, it may finally become so excessive that the river cuts through the narrow neck of the meander, so establishing a new channel that is straighter. The meander loop that has been abandoned transforms into an oxbow lake as sediments progressively fill it up over the course of time. Examples: Oxbow lakes along the Mississippi River (USA), Yellow River (China).

Figure 34.20: Diagram showing Braided Channel

7. **Braided Channels:** The network of braided channels is made up of a series of river channels that are intertwined with one another and are divided by braid bars, which are tiny islands of sediment that are often transient. The formation of braided channels occurs in rivers that have significant sediment loads and varying discharge frequencies. A portion of the sediment is deposited in the channel by the river as it reaches its maximum capacity to convey material, which results in the formation of bars. Following this, the water partitions into a number of channels that wind their way around these barriers.
8. **Crevasse Splays:** A crevasse splay is a fan-shaped deposit of sediment that is formed when a river breaches its levee and dumps silt over the floodplain. This kind of deposit is characterized by its fan-like structure.When a river breaches through its levee during a flood, the abrupt release of water and sediment spreads out in a fan-like configuration across the floodplain, generating a crevasse splay. Crevasse splays are generated when a river bursts through its levee.

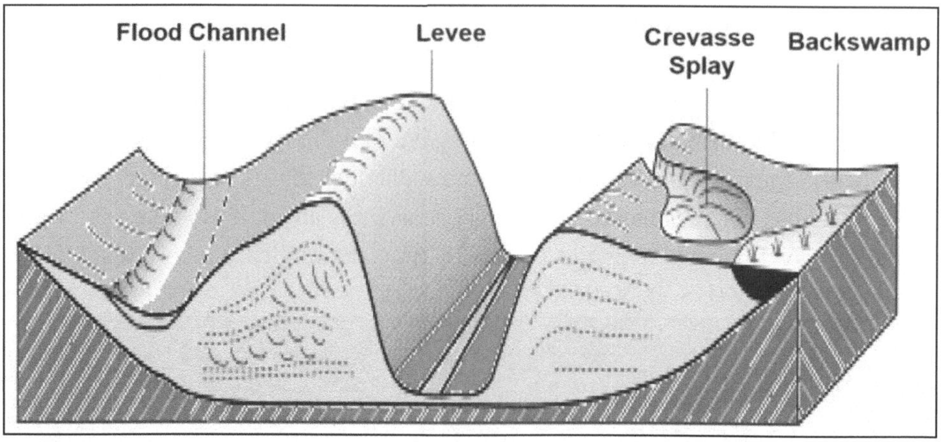

Figure 34.21: Diagram of formation of Crevasse Splays

9. **Sloughs:** On the floodplains of meandering rivers, one may find sloughs, which are depressions or channels. The formation of these landforms often takes place during flood events, when the river exceeds its banks. The water that is overflowing may carve out a quicker and alternate course over the floodplain in order to choose the path that presents the least amount of resistance. This process has the potential to create elongated and shallow depressions or channels known as sloughs. After the floods have subsided, sloughs often continue to exist as low-lying places that are filled with water. As a result of their ability to provide spaces for water storage and biodiversity within the floodplain environment, they have the potential to serve as essential habitats for a wide variety of aquatic and terrestrial species.

Figure 34.22: diagram showing a sloughs elongated, shallow depressions or channels

Other features and typical landforms associated with river and wind

1. **Pediments:** Pediments are rock-strewn surfaces that have a gentle slope and may be found near the foot of mountains in dry climates. It is essentially the process of erosion and weathering of mountains that is responsible for their formation. Rivers and streams are also contributors to the movement of material that has been eroded. As a result of the combined effects of wind and water erosion, these surfaces eventually become smooth over the course of time.

Figure 34.23: Dome-shaped Bornhardts

2. **Bornhardts:** Rock formations or hills that are enormous, dome-shaped, and solitary are known as bornhardts. Bornhardts are often comprised of hard rock such as granite. Typically, they are produced by a process known as differential weathering and erosion, in which the more resistant rock is left standing after the surrounding material, which is composed of softer material, is worn away by wind and water.
3. **Playas:** Playa is a flat-bottom depression that can be found in interior desert basins and adjacent to coasts in arid and semiarid regions. Playas are also known as pans, flat lakes, or dry lakes. Playas are occasionally covered by water that slowly filters into the ground water system or evaporates into the atmosphere, resulting in the deposition of salt, sand, and mud along the bottom of the depression and around its edges. Playas are also known as dry lakes at times.In deserts that are made up of basins that are surrounded by mountain ranges, the drainage comes from all of the basin's boundaries and flows towards the center of the basin. Playa is the name of the

large, shallow lake that emerges across this plain when there is an adequate amount of water.

Figure 34.24: Playa a flat-bottom depression found in interior desert basins

4. **Wadies:** Wadies are dry riverbeds or valleys that are prevalent in desert locations and only transport water when there is a major rainstorm on occasion. The movement of water during flash floods, which may be considerable in desert locations where rainfall is intermittent but severe, is responsible for the formation of these structures.

Figure 34.35: Dry riverbeds, Wadies

5. **Inselbergs:** Inselbergs are hills or mountains that are separated from the rest of the landscape and rise suddenly from a hilly or flat plain. They are relics of an old landscape that were produced by the erosion of the surrounding softer material, leaving behind the rock that is more resistant to erosion. Erosion caused by wind and water both have a role in the creation of inselbergs.

Figure 34.36: Inselberg, isolated hills

Important Terminologies Associated with the Actions of the Geomorphic Agent:

Effluent River: A river that absorbs more water from the groundwater as it runs is referred to as an effluent river, also known as a gaining river. This often takes place when the river channel is located below the water table, which results in groundwater seeping into the river.

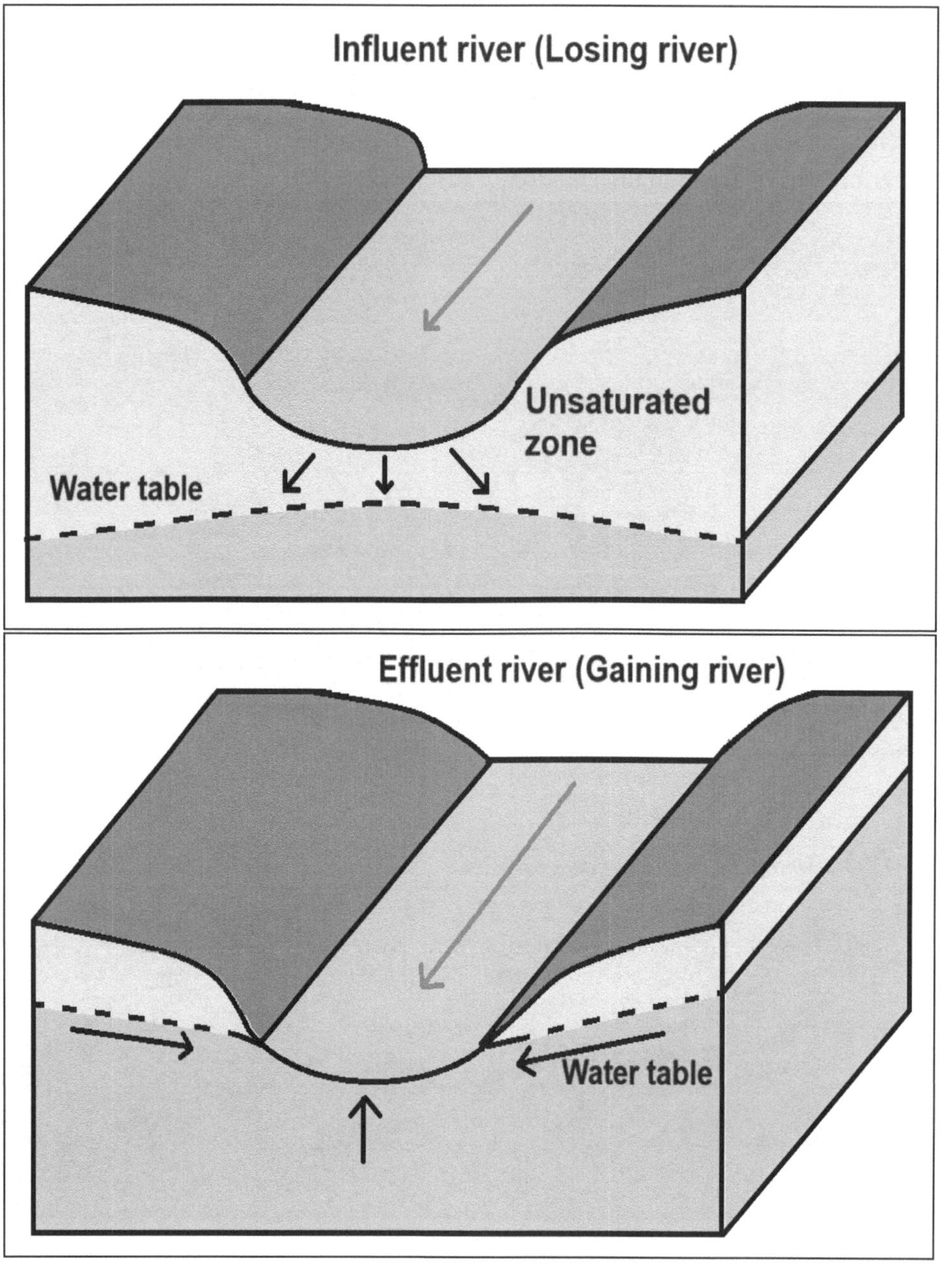

Figure 34.37: Influent River and Effluent River.

Key Concepts:

A. **Water Table**: The point below which the earth is completely saturated with water. In the case of a river that discharges water, the water table is higher than the riverbed.

B. **Baseflow**: The part of the river's flow that originates from groundwater discharge, which helps to keep the river flowing even when there is a lack of water in the ground.

C. **Perennial Flow**: As a result of steady groundwater supplies, effluent rivers often have a flow that is continuous throughout the whole year.

D. **Humid Regions**: In areas that get a lot of precipitation and have high water tables, such as those that are located in temperate and tropical climates, effluent rivers are rather prevalent.

Influent River: An influent river, one that is sometimes referred to as a losing river, is a river that, as it runs, loses water to the ground in its immediate vicinity. When the river is located above the water table, this phenomenon often takes place, which results in water seeping from the river into the ground.

Key Concepts:

A. **Recharge**: The process by which groundwater is restored by water leaking into the earth from the influent river.

B. **Ephemeral Flow**: Influent Rivers may flow seasonally or intermittently, especially in dry or semi-arid areas where the water table is often lower than the riverbed.

C. **Arid Regions**: Influent Rivers are typical in arid regions with low groundwater levels and significant rates of evaporation.

D. **Streambed Infiltration**: The method by which river water seeps into the earth and helps replenish groundwater.

Types of rivers

1. **Antecedent River:** A river is considered to be an antecedent river if it continues to follow the same path and pattern it did in the past, despite changes in the underlying rock structure or increases in the elevation of the land. A geological structure that the river passes through is older than the river itself, which is the reason why this happens. The river continues to erode its course downhill, cutting through the region that has been raised as a result of tectonic activity, particularly when the terrain is rising or becoming deformed.
2. **Consequent River:** A river is said to be a consequent river if it flows in a direction that is parallel to the slope of the ground when it first starts to flow. The drainage pattern that is developed on freshly raised ground or an area of new land surface is often the location where it develops directly. It is common for these rivers to be lined with the overall slope of the land and to follow the natural contour of the ground.
3. **Subsequent River:** The River that merges with the consequent river arising later as erosion proceeds. After the original drainage pattern has been formed, a subsequent river is one that develops its path by eroding weaker rock. This occurs after the initial drainage pattern has been established. When these rivers originate, they often do so

along areas of structural weakness, such as fault lines or softer rock beds. As they continue to exploit these weaknesses throughout time, they may shift their course.

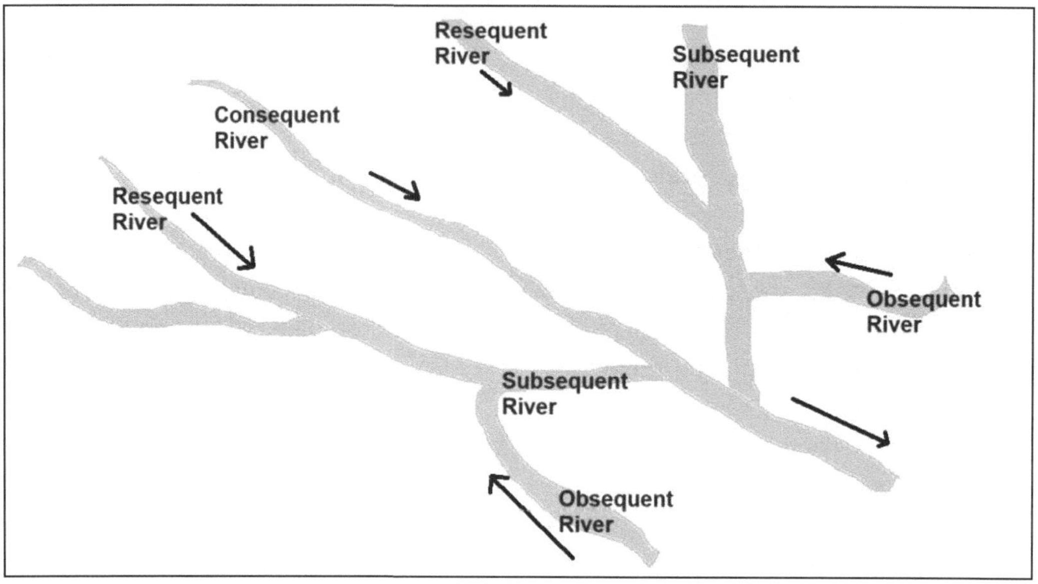

Figure 34.38: Types of rivers

4. **Insequent River:** An insequent river is a type of river that does not follow any predictable pattern of topography or structure. It appears to wander irregularly across the landscape without any obvious relation to the underlying rock structures or initial slopes. These rivers often develop in regions where the topography does not strongly control drainage patterns.
5. **Obsequent River:** The term "Obsequent River" refers to a river that flows in the opposite direction of the drainage that was responsible for its formation. This may take place as a consequence of the reversal of the slope of the land or as a consequence of complicated geological processes, such as folding or faulting, which alter the structure of the drainage pattern that was initially there.
6. **Resequent River:** Unlike the initial consequent stream, a subsequent river emerges at a lower level and often follows a different kind of rock or structural feature. It runs in the same general direction as the original consequent stream. It is possible to think of it as a subsequent river that continues to flow in a direction that is comparable to that of the first consequent stream, but has altered its path in order to take advantage of weaker rock strata or fault zones.
7. **Super-imposed River:** A river that at one time built its course on a higher level of land and then subsequently pierced through the underlying structure after the top layers were worn away is referred to as a **super-imposed river**. Because the river was first created on a different geological surface, it seems to be "draped" across the

terrain and cuts through rocks that are resistive to its flow. This causes it to ignore the structure that lies under the surface.

These concepts are helpful in gaining an understanding of the ways in which rivers interact with the geological characteristics and historical aspects of the places through which they flow.

Drainage pattern

The layout or arrangement of rivers, streams, and other drainage channels in a certain region is referred to as a drainage pattern. This pattern is formed by the topography, geology, and other environmental elements that are characteristic of the area. A great deal of information about the underlying rock structure, the different kinds of soil, and the past processes that produced the landscape may be learned from the pattern.

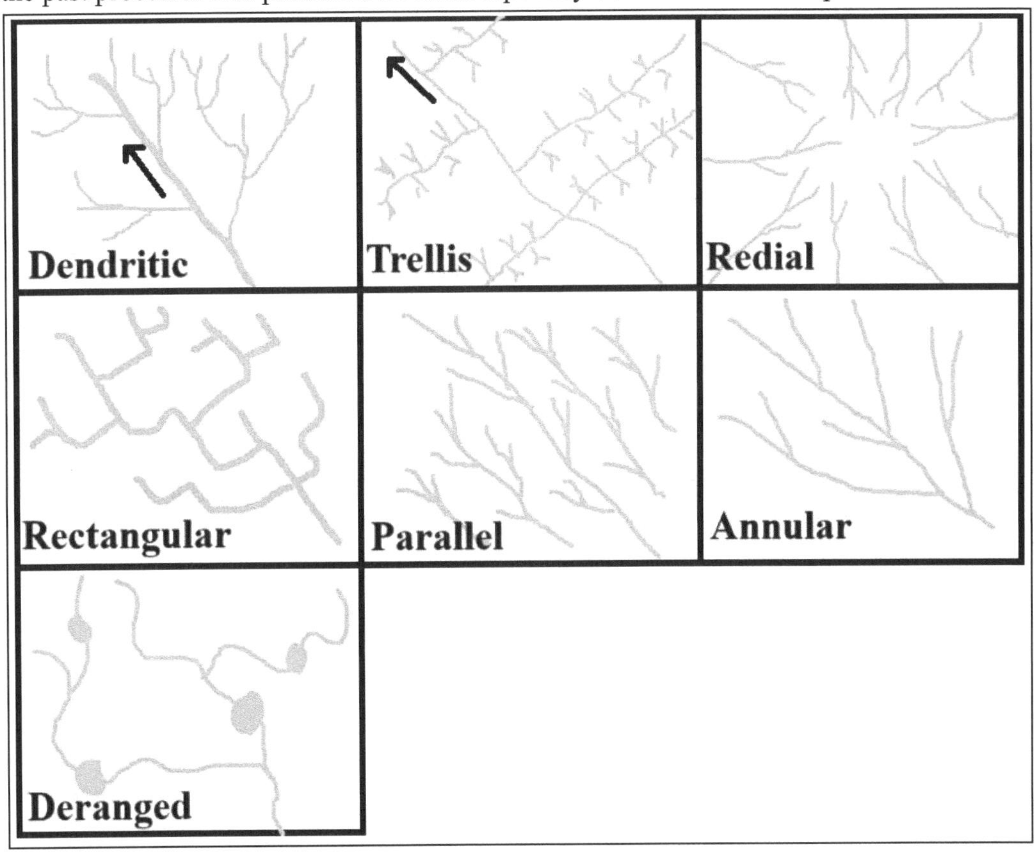

Figure 34.39: types of river drainage pattern

Common Types of Drainage Patterns

1. **Dendritic Drainage Pattern**:

A. Appearance: It resembles the branches of a tree.

B. Formation: This phenomenon takes place in areas where the rock type is homogeneous and flat-lying, which enables streams to easily branch out in all directions.
C. Example: It is typical in areas where sedimentary strata are arranged horizontally.

2. Trellis Drainage Pattern:

A. Appearance: This structure is reminiscent of a garden trellis, with major streams that are parallel to one another and shorter tributaries that connect at approximately right angles.
B. Formation: This kind of rock is often found in folded terrains and may be found in locations that have alternating bands of resistant and less resistant rock.
C. Example: Appalachian Mountains in the eastern United States.

3. Radial Drainage Pattern:

A. Appearance: A central high point serves as the hub from which streams radiate outward, much like spokes on a wheel.
B. Formation: Most often, it forms in the vicinity of volcanic cones or domes.
C. Example: Around volcanic mountains like Mount Vesuvius.

4. Rectangular Drainage Pattern:

A. Appearance: There is a rectangular grid pattern formed by the channels.
B. Formation: It is found in areas that have a fault or joint system that is well defined and that directs the flow of the stream.
C. Example: Found in areas that have a bedrock that is fractured, such as in several municipalities in the state of New York.

5. Parallel Drainage Pattern:

A. Appearance: Parallel lines are formed by the streams.
B. Formation: This kind of formation occurs in areas that have a steep slope or when the bedrock is inclined in a consistent manner.
C. Example: Frequently seen in regions that have steep slopes, such as the slopes of mountain ranges that have recently been elevated.

6. Annular Drainage Pattern:

A. Appearance: There is a centrally elevated location that is surrounded by streams that create concentric rings.
B. Formation: In most cases, it develops around dome-shaped formations that include layers of rock that alternate between being hard and soft.
C. Example: Located in the vicinity of a few eroded domes and basins

7. Deranged Drainage Pattern:

A. Appearance: The flow of the stream is erratic and chaotic, and there is no discernible pattern.
B. Formation: Glaciers and volcanic activity are two examples of recent disturbances that might cause this phenomenon to occur in certain regions.

C. Example: In areas that have seen glacial activity in recent times, such as some locations in Canada.

Each drainage pattern is a reaction to the specific geological and environmental circumstances of the region, and it has the potential to give vital insights about the history of the landscape as well as structural characteristics of the landscape.

Human Impact on Fluvial Processes and Landforms

A. River Engineering

1. Dams and Reservoirs:

- Purpose: To store water, to generate hydroelectric power, and to manage flooding.
- Impact: Transforms the transit of sediment, as well as the patterns of erosion and deposition farther downstream.

2. Channelization:

- Purpose: Control of flooding and enhancement of navigation are the goals.
- Impact: This has an effect on ecosystems, as well as increases river velocity and decreases natural meandering.

3. Levee Construction:

- Purpose: The purpose of this is to protect against floods.
- Impact: Changes the natural dynamics of the floodplain, which in turn raises the danger of flooding farther downstream.

B. Land Use Changes

1. Deforestation: Deforestation causes an increase in runoff and erosion while simultaneously reducing infiltration.
2. Urbanization: Greater urbanization results in an increase in impermeable surfaces, which in turn leads to greater runoff, less infiltration, and an increased danger of flooding.
3. Agriculture: Agriculture results in changes to the sediment load as a result of soil erosion and alters the dynamics of rivers via the use of irrigation methods.

C. Climate Change

1. Changing Precipitation Patterns: The river flow regimes are altered, and the rates of erosion and deposition are affected.
2. Glacial: Increases the amount of sediment load and river flow that originates from glacial zones.
3. Melt Sea-Level Rise: Having an impact on the creation and stability of deltas, as well as increasing saltwater incursion.

Case Studies of Fluvial Landforms

A. The Nile River

1. Floodplain and Delta Formation: Annual flooding has traditionally resulted in the deposition of nutrient-rich silt, which has led to the formation of floodplains and deltas consisting of fertile agricultural land.

2. Human Impact: The construction of the Aswan High Dam caused changes in silt deposition and decreased the fertility of the land on the downstream side.

B. The Mississippi River

1. Meanders and Oxbow Lakes: The lower Mississippi River is characterized by its extensive meandering features and the development of oxbow lakes.
2. Human Impact: The natural dynamics of the river have been drastically changed as a result of channelization efforts and levee construction.

C. The Amazon River

1. Vast Floodplains: Seasonal flooding results in the formation of abundant floodplains that are home to a wide variety of habitats.
2. Sediment Transport: A vast delta is formed as a result of sediment transport, which is responsible for transporting a tremendous load of material from the Andes to the Atlantic Ocean.

D. The Ganges-Brahmaputra Delta

1. Delta Formation: One of the largest deltas in the world, formed by the deposition of sediments from the Ganges and Brahmaputra rivers.
2. Human Impact: The Ganges and Brahmaputra rivers are responsible for the formation of one of the greatest deltas in the world, which was developed by the deposition of sediments.

Management and Conservation of River Systems

A. Sustainable River Management

1. Integrated Water Resources Management (IWRM)

a. Approach: Environmental, social, and economic considerations are all taken into account in holistic management.

b. Goals: Water resources should be used in a sustainable manner, with human demands and ecological health being balanced.

2. River Restoration

a. Purpose: The purpose of this project is to restore natural habitats and processes.

b. Methods: The removal of dams, the restructuring of waterways, and the restoration of floodplains are the methods.

B. Flood Management

1. Non-Structural Measures

a. Methods: Methods include zoning in floodplains, early warning systems, and raising awareness within the community.

b. Benefits: Reduces the danger of flooding and the damage it causes without affecting the dynamics of the river.

2. Structural Measures

a. Methods: Some of the methods include floodwalls, levees, and reservoirs.
b. Considerations: Striking a balance between providing protections and preserving the natural functioning of rivers.

C. Sediment Management

1. Dredging

a. Purpose: Maintaining navigation channels and preventing siltation.
b. Impact: The impact is that it has the potential to disturb ecosystems and calls for careful planning and control.

2. Sediment Trapping

a. Methods: Constructing sediment traps and controlling land use upstream are the methods that are being used.
b. Benefits: The benefits include a reduction in the sediment load downstream and protection of water quality.

D. Ecosystem Protection

1. Riparian Buffers

a. Purpose: The purpose of this project is to protect riverbanks, filter pollutants, and create habitat.
b. Methods: Conserving natural riparian zones and planting plants in the area.

2. Biodiversity Conservation

a. Approach: The approach consists of preserving and restoring environments, as well as controlling invasive species.
b. Goals: The goals are to preserve ecological integrity and to provide support for a wide variety of species.

Rivers are significant agents of geological change, changing landscapes via processes of erosion, movement, and deposition. These activities result in the formation of a wide variety of landforms, which are a reflection of the dynamic interaction that occurs between water and the surface of the Earth. These landforms include meandering rivers, floodplains, deltas, and gorges. River systems are still being impacted by human activities and climate change, which calls for sustainable management and conservation measures to be made in order to strike a balance between the requirements of humans and the health of the environment. It is crucial to have a solid understanding of the geological work that rivers do in order to accurately forecast future changes, effectively manage water supplies, and safeguard the critical ecosystems that are maintained by these dynamic fluvial systems.

Chapter 35
Geological Work of Glaciers and Their Landforms

Introduction to Glacial Processes

Glaciers are strong agents of geological change that modify landscapes via processes of erosion, movement, and deposition. It is through these activities that glaciers shape landscapes. Glaciers, which now cover around ten percent of the global land area, have had a huge influence on the geography of the globe, particularly during the ice ages that have occurred in the past. This chapter examines the mechanisms that are responsible for glacial processes as well as the many landforms that are produced as a result of the geological work that glaciers do.

Glacial Erosion Processes: Glacial erosion is the process by which glaciers remove and transport rock material, which ultimately results in the formation of a variety of landforms.

Types of Glacial Erosion

1. **Plucking:**"Plucking" is a term that describes the process by which a glacier displaces the bedrock that is underneath it. Because of the movement of the glacier, it freezes onto the bedrock. It is possible to say that the glacier "plucks" or pulls chunks of rock away from the bedrock as it continues to move. This happens because the underlying rocks are subjected to a considerable force from the weight and movement of the

glacier. Usually, the presence of water facilitates the plucking process. Glacier meltwater penetrates into bedrock fissures, freezes, and expands, weakening the rock. These fragile rock fragments are then dragged away by the glacier's movement, creating a rugged, ragged terrain. Glacial valleys and other landforms are mostly shaped by this process in conjunction with abrasion, which occurs when boulders buried in the glacier grind against the bedrock.

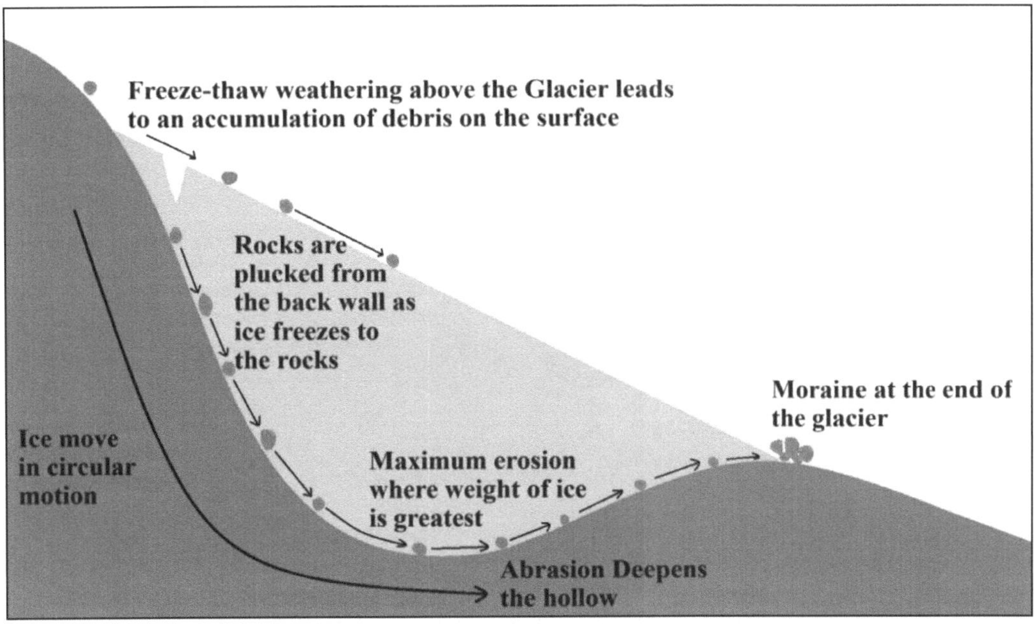

Figure 35.1: Glacial erosion by Plucking and Abrasion

2. **Abrasion**:"Abrasion" is the process by which debris and boulders lodged in a glacier's bottom rub up against the bedrock below it. As the glacier continues to move forward, the rocks that are embedded inside it perform the function of sandpaper, polishing and smoothing the surface of the bedrock. One of the key methods by which glaciers erode the terrain, carving out valleys, fjords, and other glacial landforms, is abrasion. Puckering is another technique that glaciers use to erode the landscape. The unique U-shaped valleys that are often associated with glacial activity are the consequence of the collision of two processes. There are several ways in which the process of abrasion might take place:

a. **Scouring:** The movement of the glacier causes the boulders and silt that it carries to scour the bedrock, which wears it down and smoothes it out. This process is known as scouring.

b. **Striations:** As rocks embedded in the glacier drag across the bedrock, they can leave behind long, parallel grooves or scratches known as striations. These striations provide evidence of the direction of the glacier's movement.

c. **Polishing:** The bedrock surface may be polished by using finer particles, such as sand or silt, which will leave it smooth and shining after the polishing process.

3. **Avalanching:** Avalanching is a process that occurs when significant volumes of snow, ice, and debris flow down a hillside in a quick manner. This movement is often caused by gravitational forces.

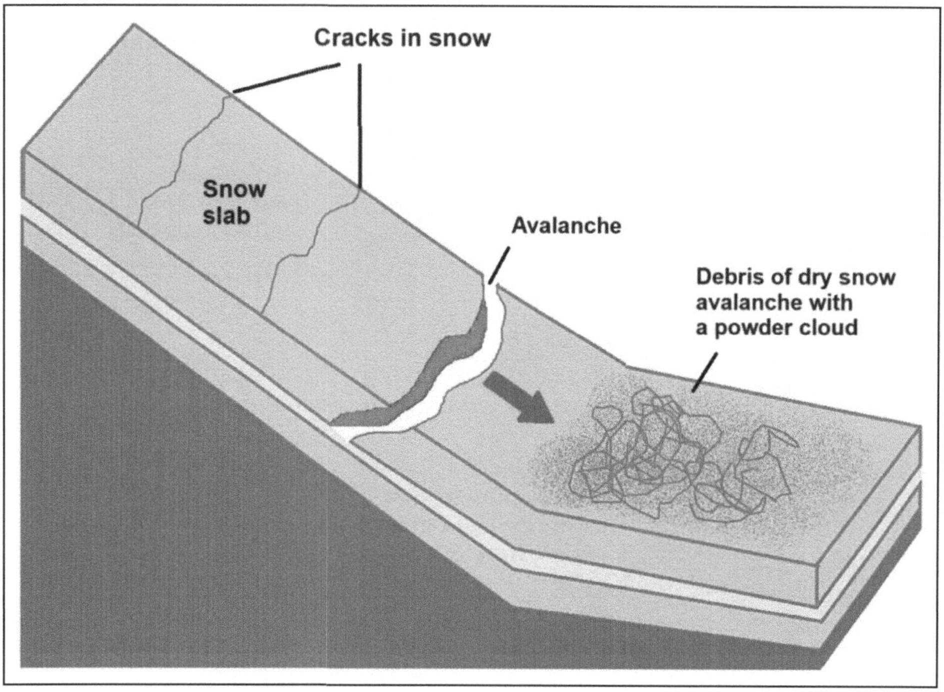

Figure 35.2: Glacial erosion by Avalanching

4. **Nivation:** The process of nivation is characterized by the weathering and erosion of the ground under and surrounding a snow patch. This process is driven by the repetitive freeze-thaw cycles that occur throughout the year. In particular, it plays a significant influence in the formation of alpine landscapes, especially in high-latitude or high-altitude places where snow patches remain for lengthy periods of time or exist throughout the whole year.

5. **Ablation:** The term "ablation" refers to the process by which ice and snow are removed from a glacier or ice sheet by melting, sublimation, calving, or evaporation depending on the circumstances. It is possible for ablation to take place on the surface of the glacier, along its margins, or at its base. Ablation is an essential component in the process of mass balance in glaciers.

This process can contribute to glacial erosion in several ways:

A. **Mass-Wasting Process**: In the process of avalanching, this is classified as a mass-wasting process, significant quantities of rock, debris, and other materials slide downslope under the impact of gravity. This phenomenon often takes place along the steep walls of valleys that have been carved out by glaciers in glacial settings.

B. **Valley Margins and Erosion**: When glaciers travel through valleys, the sheer weight and movement of the ice may force the valley sides to scrape against the glacier, which can result in the granite blocks breaking off. This can happen when glaciers move through valleys. These fractured chunks have the potential to get entrenched in the glacier or to freeze there.

C. **Undercutting and Slumping**: The glacier maintains it's scraping and erosion along the valley borders, which ultimately results in the undercutting of the valley walls. The slopes become unstable as a result of this, which encourages slumping, which is the downward sliding of rock and soil, as well as sliding, which further adds to the mass of debris that avalanches down.

D. **Debris Avalanching**: An avalanching of debris is the consequence of slumping and sliding processes, which are made worse by undercutting. This results in the glacier's surface being covered with a significant amount of rock and other debris.

E. **Surface Load of the Glacier**: Avalanching is responsible for bringing rock debris onto the surface of the glacier, which contributes to the glacier's burden. This material is carried along with the ice as the glacier moves, which contributes to the erosion of the landscape as the ice that is loaded with debris grinds against the bedrock.

This process not only results in the construction of the valley that the glacier travels through, but it also adds to the general modification of the landscape in areas that experienced glaciations. The debris that is carried by the glacier plays a key part in the process of shaping the landscape. It helps in the construction of features such as moraines and contributes to the deposition of glacial till.

A. Factors Influencing Glacial Erosion

1. Glacial Velocity: Glaciers that are moving faster wear away more quickly because of the greater friction and pressure they experience.
2. Ice Thickness: If the ice is thicker, it will impose greater pressure, which will increase both plucking and abrasion.
3. Rock Type: Rock that is softer is more prone to erosion than rock that is tougher.
4. Debris Load: The quantity of debris carried by the glacier, as well as the size of the debris, has an effect on the severity of the abrasion.

Glacial Erosional Landforms

1. **U-Shaped Valleys:** Glaciers and their erosive effect, which deepens and widens river valleys that already exist, are responsible for the formation of U-shaped valleys. A U-shaped cross-section is characteristic of glacial valleys, in contrast to the V-shaped valleys that are formed by rivers. Glacial valleys feature high, straight sides and a flat

bottom of the valley. Examples: Yosemite Valley (USA), Lauterbrunnen Valley (Switzerland).

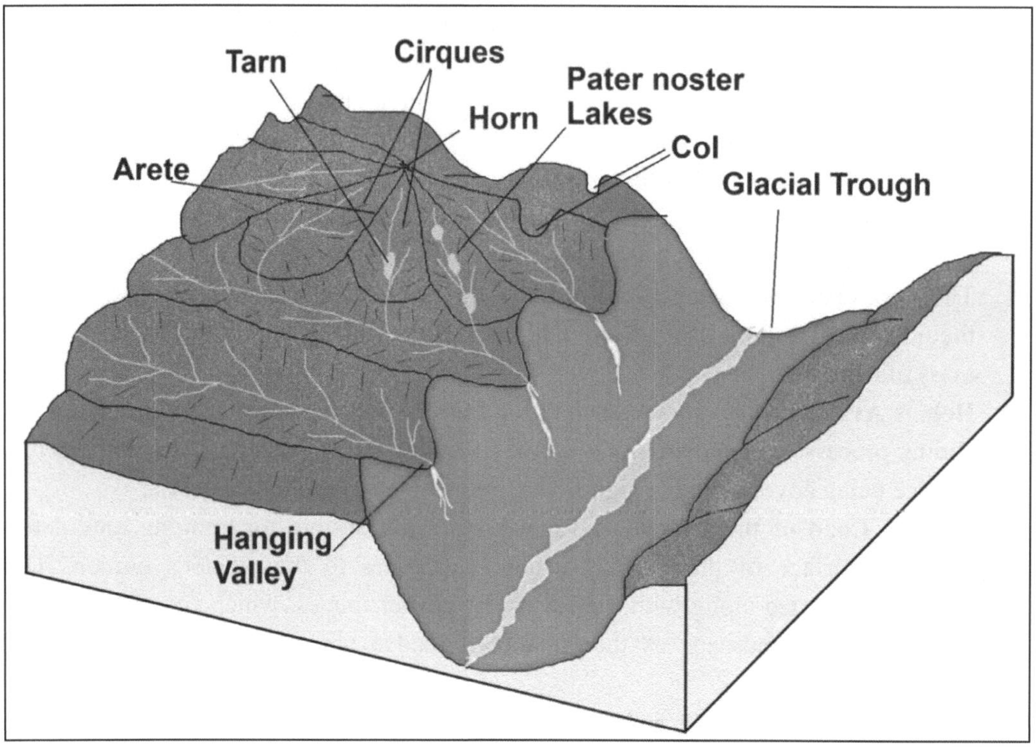

Figure 35.3: Glacial Erosional Landforms

2. **Cirques:** A cirque is a dip at the top of a glacier valley that resembles an amphitheater and is bowl-shaped. A glacier's erosive process, which creates a depression in the slope over time as it builds up snow and ice, creates cirques. Examples: Cirques in the Alps and the Scottish Highlands.
3. **Tarn:** A tarn is a tiny lake or pool that is often located in a cirque and is found in mountainous regions. It comes into existence as a result of the glacier melting and fills the depression with water. Example:LlynIdwal in Snowdonia, Wales.
4. **Arêtes:** Arêtes are ridges that are steep and narrow, and they are formed between two glacial valleys or cirques that are next to one another. The formation of these features is a result of the headward erosion of two glaciers that are moving in the opposite direction**.** Example: The Knife Edge on Mount Katahdin, Maine, USA.
5. **Horns**: The erosion of at least three glaciers that meet at a single place results in the formation of a horn, which is a pointed peak in the shape of a pyramid by the glacier. Every glacier erodes the mountain from a different side, which results in the top becoming more pointed. Examples: Matterhorn (Switzerland), Mount Assiniboine (Canada).

6. **Col:** The term "col" refers to a low point or a pass that divides two mountain summits. A saddle-like depression is created as a result of the erosion of the ridge that separates two glaciers that are in opposition to one another. It is common practice to utilize cols as mountain passes.
7. **Fiords:** Fjords, also known as fiords, are inlets of the sea that are deep and narrow, and they are produced when a glacier valley is flooded. These inlets are located between high cliffs or steep slopes. Geographically speaking, they are often located in areas that were glaciated in the past but have since been submerged as a result of increasing sea levels. **Example:** Sognefjord in Norway.

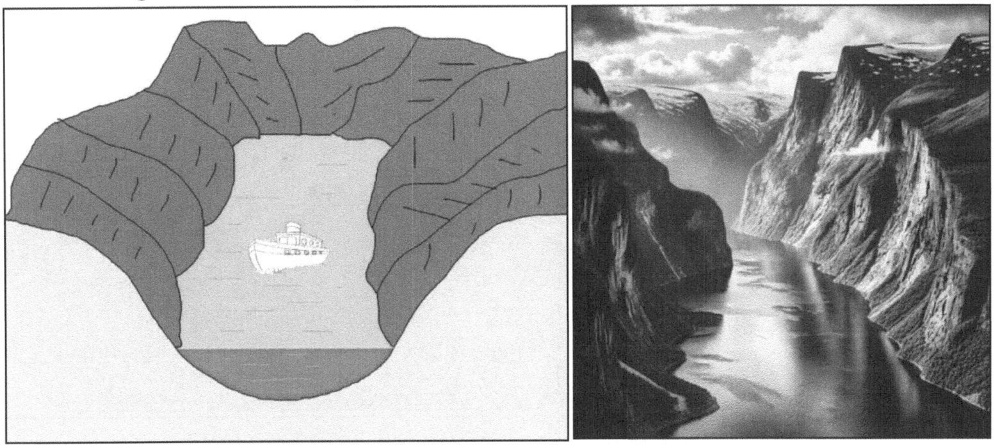

Figure 35.4: The image of a stunning fjord inspired by the Sognefjord in Norway.

8. **Striations:** The movement of a glacier may leave scratches or grooves on the surface of rocks, which are referred to as striae. These markings are often linear, and they may be used to determine the direction in which the glacier is moving. They come into existence as a result of the glacier dragging boulders and debris over the bedrock.

Figure 35.5: Striationson rocks by glacier's movement.

9. **Roche Moutonnée:** Roche Moutonnée is asymmetrical rock formations that were formed as a result of glacial abrasion on the upstream side and plucking on the downstream side. There is a side that is steep and jagged, as well as a side that is smooth and softly slopes. Examples: Found in glaciated regions worldwide.

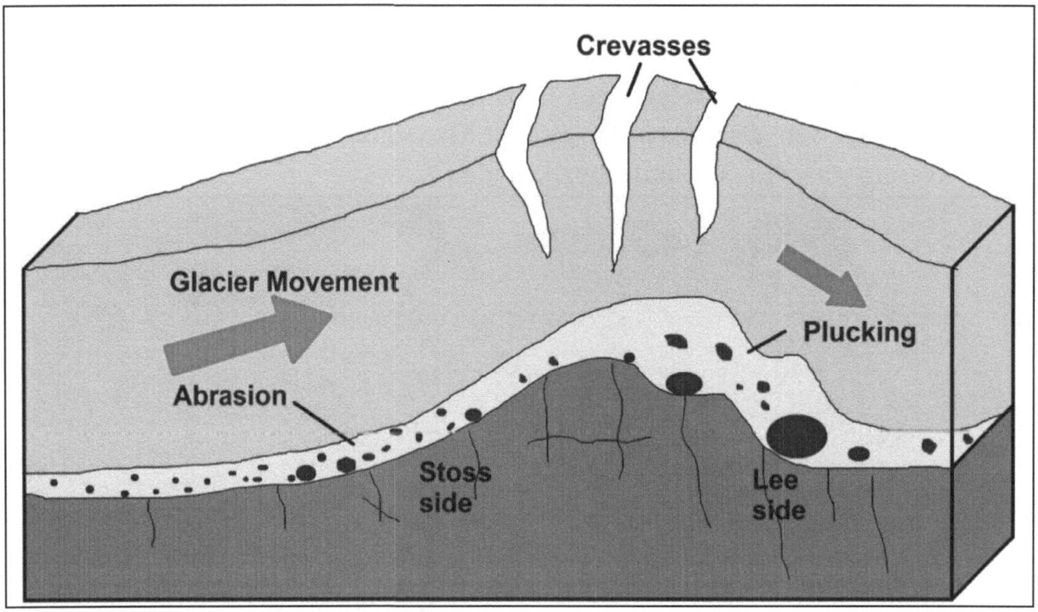

Figure 35.6: Roche Moutonnée and Crevasses: Glacial Erosional Landforms.

10. **Crevasses:** A crevasse is a deep fissure or fracture that may be discovered in a glacier. Crevasses are mainly created by the movement and tension of the ice as it flows across uneven terrain. The width and depth of these structures may vary, with some of them reaching several meters in width and several meters in depth. For mountaineers and explorers, crevasses are perilous because they are often concealed behind snow bridges, which might collapse under their weight. Crevasses are created when the glacier passes over an obstruction or a curve, stretching and breaking the ice. Transverse (across the glacier) and longitudinal (following the glacier's flow) are the most prevalent forms. They often show up in a glacier's topmost, fragile layers, when the ice is stressed.
11. **Crag and Tail:** A landform known as a crag and tail is characterized by the presence of a resistive rock, known as the crag, which is shaped like a steep hill, and a gradual slope of softer material, known as the tail, that extends down-glacier from the crag. The tail is shielded from erosion by the crag, which is what led to the construction of this unusual feature. Example: Edinburgh Castle Rock, with the Royal Mile forming the tail, in Scotland.

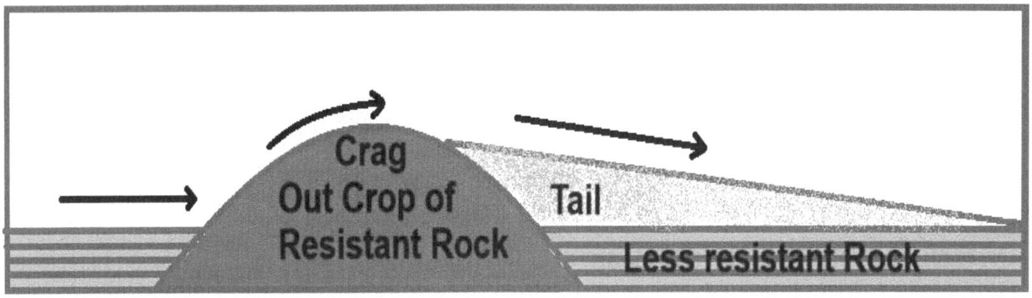

Figure 35.7: Crag and tail

12. **Serac:** The term "serac" refers to a big, unstable block or column of ice that is often discovered on the surface of glaciers, especially in crevassed regions or on the steep slopes of icefalls. The extreme fracture of the glacier ice that occurs when it flows across areas of uneven terrain is what causes seracs to emerge.

Glacial Transport Processes

There are a variety of techniques that glaciers use to convey a broad variety of materials, ranging from little particles to massive boulders.

1. **Supraglacial Transport:**

A. Process: The debris that is moved on the surface of the glacier, which is often of the origin of rockfalls and avalanches.

B. Environment: Common in mountainous regions.

2. **Englacial Transport**

A. Process: Debris that are found encased inside the ice and are transported by the glacier as it advances.

B. Environment: Throughout the whole of the glacier's main body.

3. **Subglacial Transport**

A. Process: A large amount of erosion is often caused by debris that is moved near the foot of the glacier.

B. Environment: It is located under the glacier, where it may be crushed into a fine substance.

Glacial Deposition Processes

During the process of melting and retreating, glaciers deposit the materials that they have carried, resulting in the formation of a variety of depositional ecosystems.

A. Types of Glacial Deposition

1. **Moraines**

A. **Terminal Moraines:** There are ridges of debris that have been deposited near the terminal of the glacier, which indicate the height of the glacier's progress. The ridges are curved and run perpendicular to the flow of the glacier. Examples: Long Island (USA), the Southern Alps (New Zealand).

B. **Lateral Moraines:** An accumulation of debris that runs down the slopes of a glacier. It is ridges that are parallel to the flow of the glacier. Examples: Found along the sides of many alpine glaciers.

C. **Medial Moraines:** The formation of medial moraines occurs when two glaciers collide and their lateral moraines consolidate onto one another. An example of this would be ridges of debris that run along the middle of a glacier. Examples: Found in regions with converging glaciers.

D. **Ground Moraines:** A blanket of till that has been deposited under a glacier is known as ground moraines. It is a landscape that is hilly and rolling. Examples: Found in areas of extensive glacial retreat.

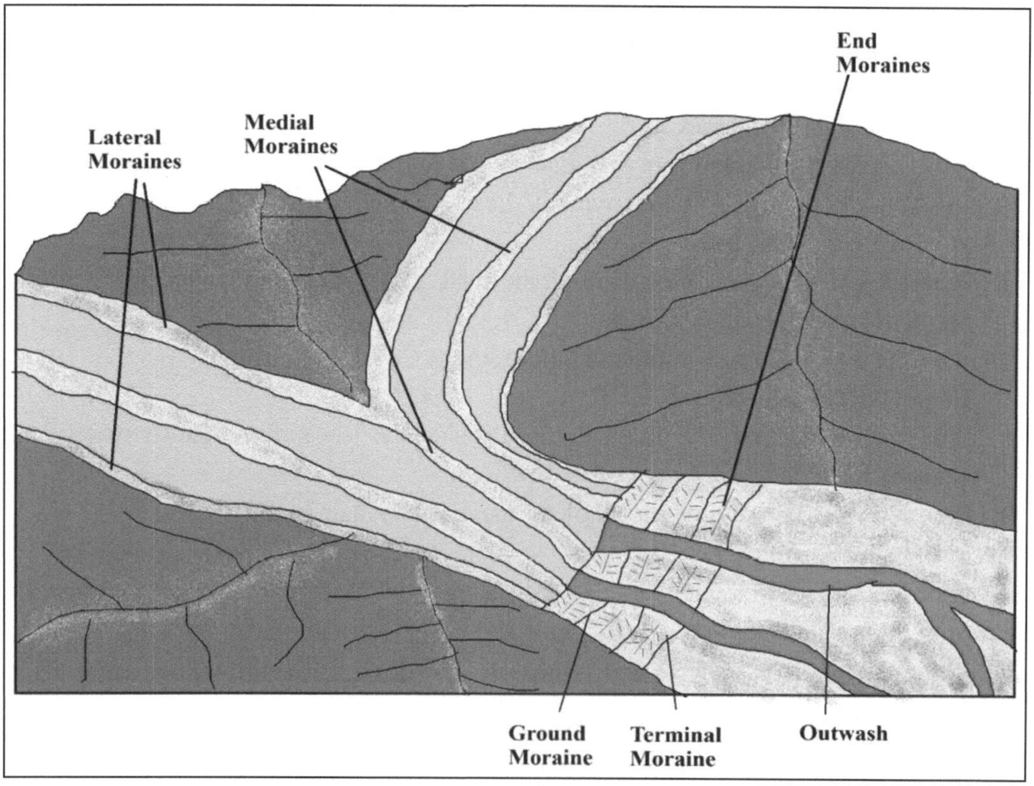

Figure 35.8: Types of Moraines

E. **Recessional moraines:** In contrast to terminal moraines, recessional moraines are generated during brief intervals in the glacier's retreat. Recessional moraines are comparable to terminal moraines. Debris is deposited on ridges that are parallel to the terminal moraine when the glacier periodically stops moving and retreats. Indicative of the glacier's gradual retreat, these moraines have formed a sequence of ridges together.

2. **Drumlins:** Drumlins are streamlined hills of terrain that are produced under a glacier and serve to indicate the direction in which ice is flowing. An elongated design with a steep stoss end and a tapering lee end give it a distinctive appearance. Examples: Drumlin fields in Ireland, the Midwest USA.
3. **Eskers:** Eskers are sweeping hills of sand and gravel that were produced by melt water streams that originated under glaciers. The ridges are quite lengthy and twisting. Examples: Found in glaciated regions, such as Canada and Scandinavia.
4. **Kames:** On the surface of the glacier or at its border, kames are mounds or hills of sand and gravel that are deposited by melt water in depressions. The terrain is erratic and often has steep slopes. Examples: Found in glacial outwash plains.
5. **Outwash Plains:** A glacier's melt water streams dump debris in vast, flat expanses known as outwash plains. These plains are characterized by their flatness. Sand and gravel are arranged in layers that are stratified. Examples: Found in front of retreating glaciers, such as in Iceland and Alaska.

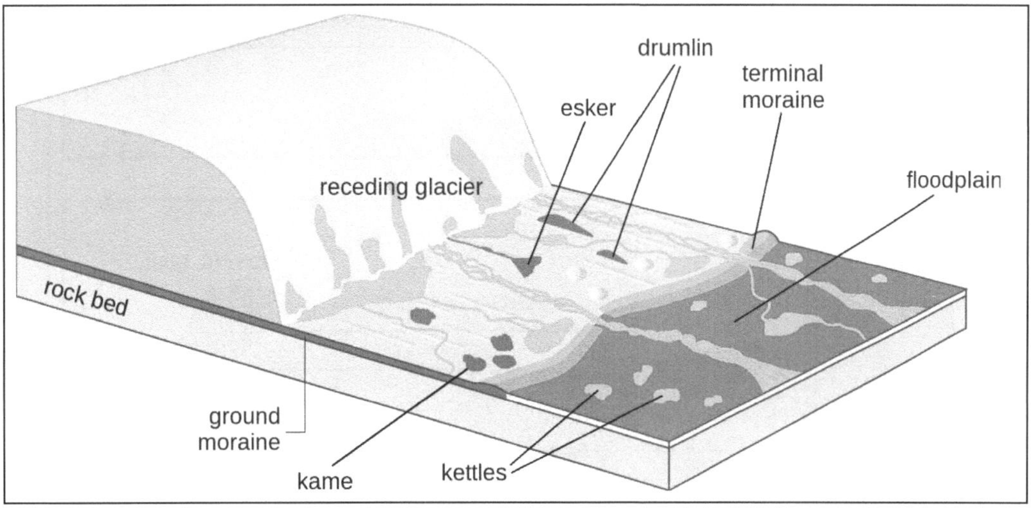

Figure 35.9: Glacial Deposition landforms

6. **Kettles:** The term "kettle" refers to a body of water that is shallow and loaded with silt. It is generated as glaciers recede or when floods drain away. It is common for glaciers to recede, leaving behind enormous chunks of ice that are buried in sedimentous deposits. The ice block will eventually melt, leaving behind a dip in the ground that is referred to as a kettle hole. It is possible for the depression to transform into a kettle lake or kettle pond if it is filled with water.
7. **Erratic:** A rock or boulder that has been moved by glacial ice and deposited, often at a location that is distant from its original location, is referred to as a glacial erratic. The composition of these rocks, which often varies from the bedrock upon which

they lay, may be used to identify them. The size of these rocks ranges from pebbles to enormous boulders.

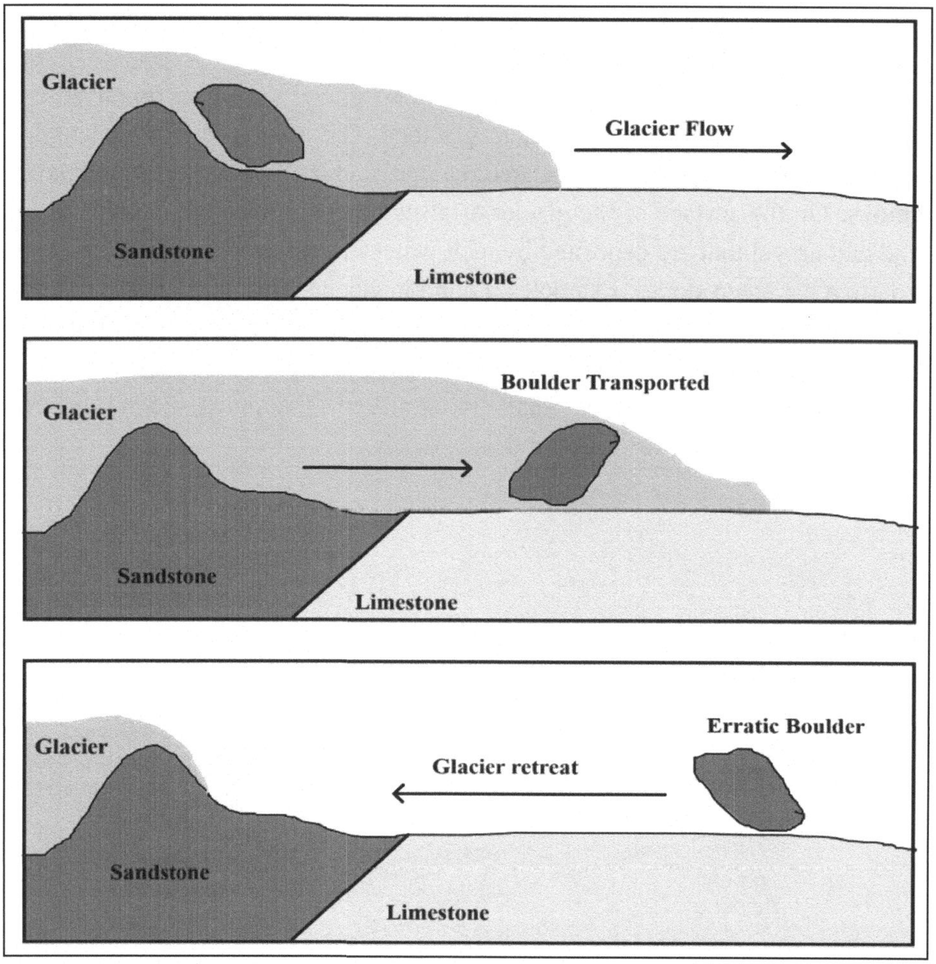

Figure 35.10: Process of transported of Erratic by glacial ice

8. **Calving:** The process of massive pieces of ice breaking off from the edge of a glacier, ice shelf, or iceberg and falling into a body of water is referred to as "calving." In most cases, the process of caving takes place at glaciers that end in the ocean or a lake, resulting in the formation of icebergs that drift away.
9. **Iceberg:** Large chunks of freshwater ice that have broken off from glaciers or ice shelves and are now floating in open sea are referred to as icebergs. Polar Regions are the most common places for icebergs to develop, and the North Atlantic and Southern Oceans are the most common places for them to be discovered.
10. **Nunatak:** A mountain, ridge, or peak that is exposed and protrudes above an ice sheet or glacier is referred to as a nunatak. Nunataks are critical habitats for life in

glaciated environments because they are often found in secluded areas and are not covered by ice.

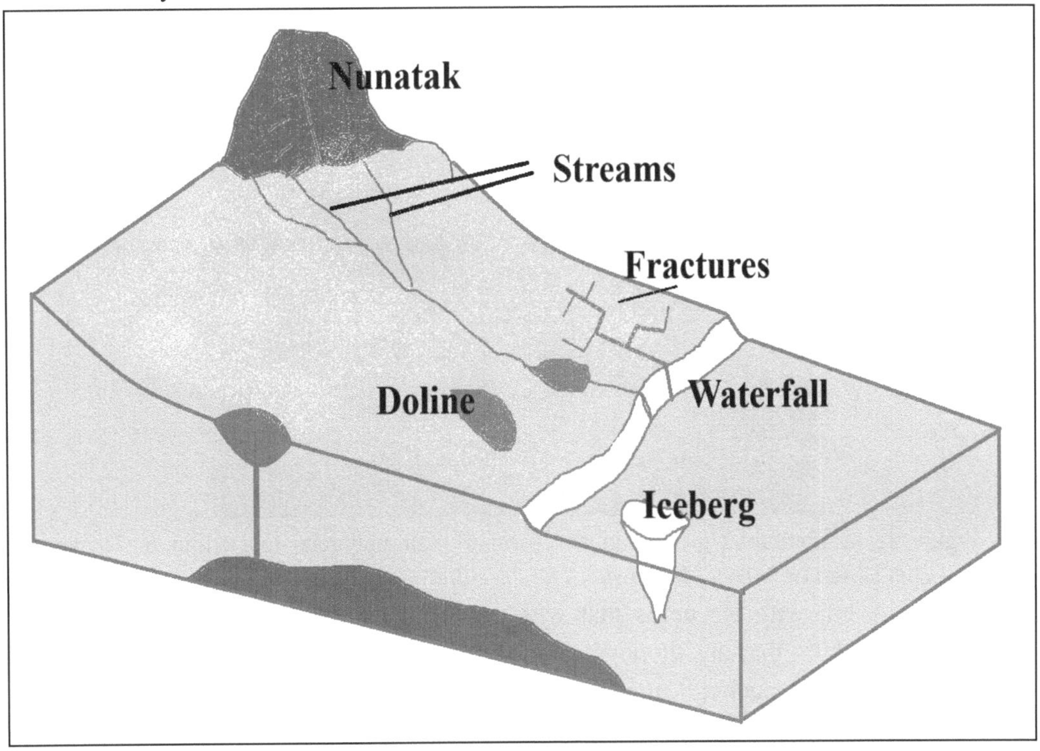

Figure 35.11: Nunatak, doline and Iceberg.

11. **Moulin:** A moulin, sometimes known as a glacier mill, is a significant glacial landform. A moulin is created when surface melt water seeps into crevasses or fissures on a glacier or ice sheet. The water erodes the ice and forms a vertical shaft; this is often the result of melting ice brought on by higher temperatures. This shaft, also known as a moulin, facilitates the movement of the glacier by allowing water to pass from its surface to its base. Moulins may vary in size from tiny apertures to massive, well-like structures that can be several meters wide and hundreds of meters deep. Moulins can also be found in a variety of shapes and sizes. It is possible for melt water to reach the base of the glacier via these conduits, which in turn may have an effect on the way the glacier slides over the rock or sediment that lies underneath it.

Periglacial Processes and Landforms

Periglacial Environment: A region that is characterized by freeze-thaw phenomena and is located next to glaciers or ice sheets. Permafrost and periodic freezing and thawing are the driving forces behind the periglacial ecosystem.

Periglacial Landforms

1. **Patterned Ground**: Patterns in the earth that are regularly spaced and geometric in nature, generated by cycles of frost and thaw. Polygons, circles, and stripes together make up this. Examples: Found in periglacial areas worldwide.

Figure 35.12: Patterned ground in the form of soil polygons located in the hyper-arid Atacama Desert (reference: *Wikipedia, The Free Encyclopedia.*)

2. **Pingo**: A hill with ice cores that was produced by the uplift of frozen earth. It consists of hills that are dome-shaped and round. Examples: Found in Arctic and subarctic regions, such as Canada and Siberia.

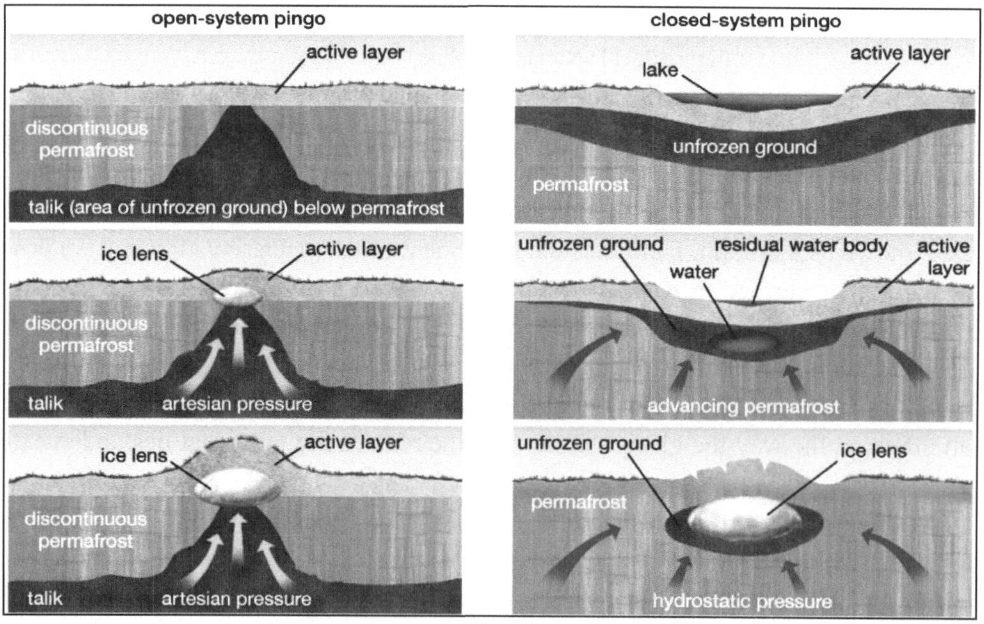

Figure 35.13: Diagram explaining the formation of open-system and closed-system pingos. (Reference: Encyclopedia Britannica).

Figure 35.14: The image of a pingo, depicting the ice-cored hill in a cold, Arctic landscape.

3. **Thermokarst**: The melting of permafrost results in the formation of a topography that is uneven and hummocky. Depressions, sinkholes, and mounds are some of the traits that it has. Examples: Common in Alaska and Siberia.

Figure 35.15: The image of a thermokarst landscape, showing the characteristic irregular terrain with depressions, sinkholes, and mounds.

4. **Solifluction Lobes**: The movement of water-saturated soil across permafrost in a gradual, downslope moving motion. The characteristics are lobate and terraced, and they are located on hills. Examples: Discovered in periglacial zones that are located at high altitudes and high latitudes.

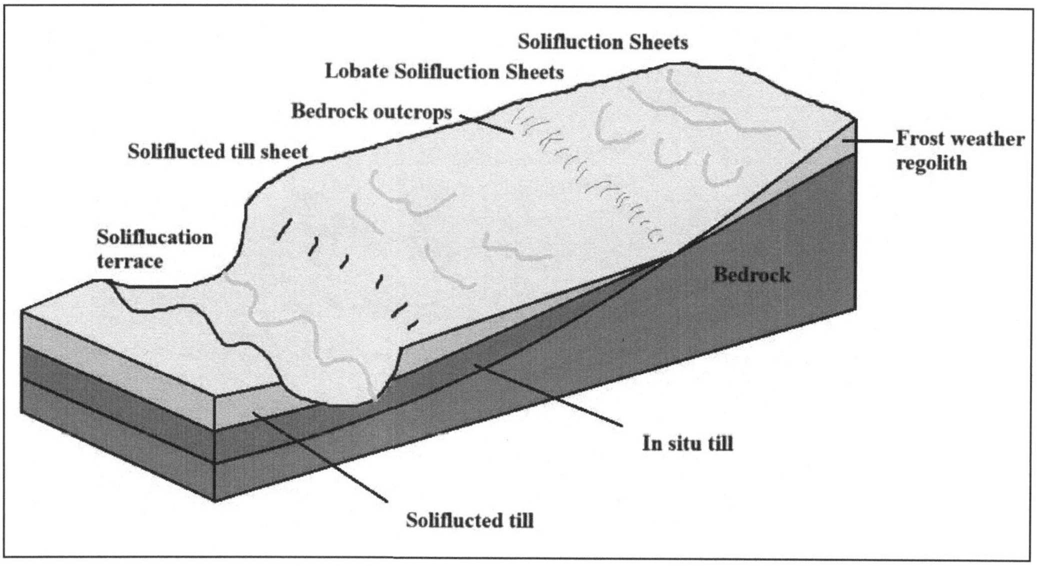

Figure 35.16: A schematic diagram of solifluction phenomena

Human Impact on Glacial Processes and Landforms

A. Climate Change

1. Glacial Retreat

a. Impact: Glaciers all across the globe are melting and disappearing at an accelerated pace.

b. Consequences: The increase in sea level, the depletion of freshwater supplies, and the alteration of glacial landscapes.

2. Permafrost Thaw

a. Impact: Permafrost is melting as a result of increasing temperatures.

b. Consequences: Greenhouse gas emissions, infrastructural damage, and thermokarst formation.

B. Tourism and Recreation

1. Infrastructure Development

a. Impact: Building infrastructure in glacial areas, including roads, resorts, and amenities.

b. Consequences: Habitat disturbance, increased erosion, and changes to natural landscapes.

2. **Recreational Activities**

a. Impact: Mountaineering, skiing and Hiking.

b. Consequences: Compaction of the soil, harm to the plants, and a rise in pollutants and garbage.

C. Resource Extraction

1. **Mining**

a. Impact: Resource and mineral extraction in periglacial and glacial regions.

b. Consequences: Pollutants, increased sedimentation, and habitat degradation.

2. **Hydropower Development**

a. Impact: Building reservoirs and dams in streams of glacier melt water.

b. Consequences: Disturbance of ecosystems, modifications to sediment movement, and changes in river dynamics.

10. Management and Conservation of Glacial Landscapes

A. Protecting Glacial Ecosystems

1. **National Parks and Reserves**

a. Purpose: Preserve ecosystems and landscapes associated with glaciations.

b. Examples: Glacier National Park (USA), Jostedalsbreen National Park (Norway).

2. **Sustainable Tourism Practices**

a. Approach: Reduce the negative effects on the environment while encouraging eco-friendly travel.

b. Methods: Trash minimization, trail maintenance, and tourist education.

B. Mitigating Climate Change Impact

1. **Reducing Greenhouse Gas Emissions**

a. Approach: Cut down on carbon emissions to help mitigate global warming.

b. Methods: Reforestation, energy efficiency, and renewable energy.

2. **Monitoring Glacial Changes**

a. Approach: Utilize field research and remote sensing to monitor glacial changes.

b. Examples: The European Space Agency's CryoSat project and NASA's Operation Ice Bridge.

C. **Adapting to Permafrost Thaw**

1. **Engineering Solutions**

a. Approach: Build infrastructure to resist melting permafrost.

b. Methods: Heat exchangers, movable supports, and insulated foundations.

2. **Community Relocation**

a. Approach: Communities should be moved from high-risk zones.

b. Examples: Villages in Alaska are moving as a result of permafrost thawing and coastline erosion.

Glaciers are powerful geological influences that have had a significant impact on the surface of the Earth via the processes of erosion, movement, and deposition that they have facilitated. It is a tribute to the dynamic character of glacial activity that the different landforms generate, which range from towering horns and magnificent U-shaped valleys to rolling drumlins and enormous outwash plains, are created by melting glaciers. These delicate ecosystems are still being influenced by human activities and climate change, which calls for the implementation of sustainable management practices and conservation initiatives in order to ensure that these amazing landscapes are preserved for future generations. For the purpose of anticipating future changes and reducing the effects of glaciers on both natural and human systems, it is essential to have a solid understanding of the geological work that glaciers do.

Chpter 36
Geological work of Underground-Water and their landforms

Karst topography is the name given to the terrain that was formed as a result of the activity of groundwater throughout geologic time. The breakdown of soluble rocks like limestone, dolomite, and gypsum, which are worn over time by the slightly acidic water from rainfall and groundwater, is the primary process that results in the formation of this one-of-a-kind sort of landscape. The development of karst topography is particularly successful when the two requirements that are needed to its development are met:

1. **Soluble Rocks Near or at the Earth's Surface:** The existence of rocks that are readily dissolved by water, such as limestone, dolomite, or gypsum, is an essential component. In order for these rocks to interact with groundwater and rainfall, which is often somewhat acidic owing to the presence of dissolved carbon dioxide, they need to be near to the surface or exposed at the surface.
2. **Dense, Highly Jointed, and Thin-Bedded Rocks:** It is necessary for the rocks to have a structure that allows water to penetrate deep into them. The joints should be well-jointed (meaning they should have cracks, fractures, or faults), and they should be sparsely bedded so that groundwater may readily find its way through. The disintegration process and the creation of karst facilitate features like caves and sinkholes. In addition to the two basic prerequisites, there are also two secondary

conditions that are favorable to the formation of karst topography. These characteristics are as follows:

a. **The Presence of a Deeply Entrenched Valley of a Master Stream:** One of the factors that might affect the flow of groundwater is a master stream, which is a major river or huge watercourse that has carved deeply into the environment. Because of the valley's gradient, groundwater is able to pass through the soluble rocks with greater activity. This speeds up the process of dissolution and contributes to the construction of karst features such as caves and subterranean streams.
b. **Moderate Amount of Rainfall:** Karst formation is especially aided by modest levels of rainfall, despite the fact that rainfall is necessary for the maintenance of groundwater systems. A lack of rainfall may not give sufficient water to adequately dissolve the rocks, while an excessive amount of rainfall may result in surface runoff rather than infiltration. Sufficient water percolates through the rocks and drives the dissolution process, which in turn shapes the landscape, when there is a reasonable and continuous amount of rainfall.

Karst cycle

The Karst Cycle is a geomorphological concept that describes the creation and change of karst landscapes over the course of time. Geomorphologists developed this model in the beginning with the intention of describing the transition of a landscape that was mostly generated by the dissolving of soluble rocks like limestone, dolomite, and gypsum. The chemical weathering that occurs as a result of groundwater impact is the primary driver of the process. The Karst Cycle may be broken down into four separate phases, each of which exhibits distinctive landforms and drainage characteristics:

1. Youthful Stage

A. Predominated by the development of subterranean drainage networks.
B. Dolines, tiny sinkholes, and vertical shafts start to form.
C. Water rapidly seeps into the ground, indicating poor surface drainage in the landscape.
D. Aggressive chemical weathering causes subsurface structures like vertical shafts and tiny caverns to begin to emerge.

2. Mature Stage

A. There are well-developed karst features, including as vast cave systems, uvalas (merged sinkholes), and enormous sinkholes.
B. Water largely travels via a subterranean network, with almost little surface drainage.
C. Large subterranean passages and caves are often found.
D. Some surface depressions may collapse or experience fluctuations in groundwater levels, resulting in the formation of ephemeral lakes.

3. **Late Maturity Stage**

A. As subterranean drainage stabilizes, active karst growth decreases.

B. Dolines and sinkholes may combine to create sizable poljes or depressions.

C. The surface features exhibit evidence of subsidence and collapsing caverns.

D. The general karst topography starts to deteriorate, and in some places, surface drainage returns.

4. Old Stage

A. When the pace of breakdown drastically drops, the karst landscape becomes subdued.

B. Deeply rooted valleys and surface streams resurface, following the routes of widening depressions or collapsed cave systems.

C. The subterranean drainage system breaks down, and sediments accumulate in the remains of sinkholes and caves that have collapsed.

D. With the majority of karst features being remnants or dormant, the terrain seems more stable.

Erosional Features

1. **Grikes:** The deep, vertical cracks or clefts that develop along the joints of limestone rocks in a pavement are known as grikes. Grikes are created by chemical weathering, which occurs when naturally occurring limestone joints and fissures are penetrated by slightly acidic rainfall that contains dissolved carbon dioxide. As a result of the reaction between the carbonic acid in the water and the calcium carbonate in the limestone, the rock is progressively dissolved along these lines of weakness. These joints are broadened and deeper throughout the course of time, which results in the formation of unique vertical grooves. The depth of a grike may range anywhere from a few centimeters to several meters, depending on the model. The look of these garments is often that of a trench coat (trench-like appearance). The surface of the limestone is divided into smaller blocks by grikes, which produce a crisscross pattern and divide the surface.
2. **Clints:** Clints are the solid blocks or ridges of limestone that are found between grikes. They have a flat surface and are solid underneath. The leftover rock masses that are left behind after the surrounding joints (grikes) have enlarged and deeper as a result of chemical weathering are referred to as clint accumulations. The clints are the areas of the limestone pavement that have been less eroded, and they stand in contrast to the grikes, which have been removed. In contrast to grikes, clints often have flat, smooth surfaces because of less chemical degradation. Depending on the grikes' spacing and pattern, they might vary in size and form. The pavement has the appearance of a blocky chessboard because the clints are spaced apart by a system of grikes.

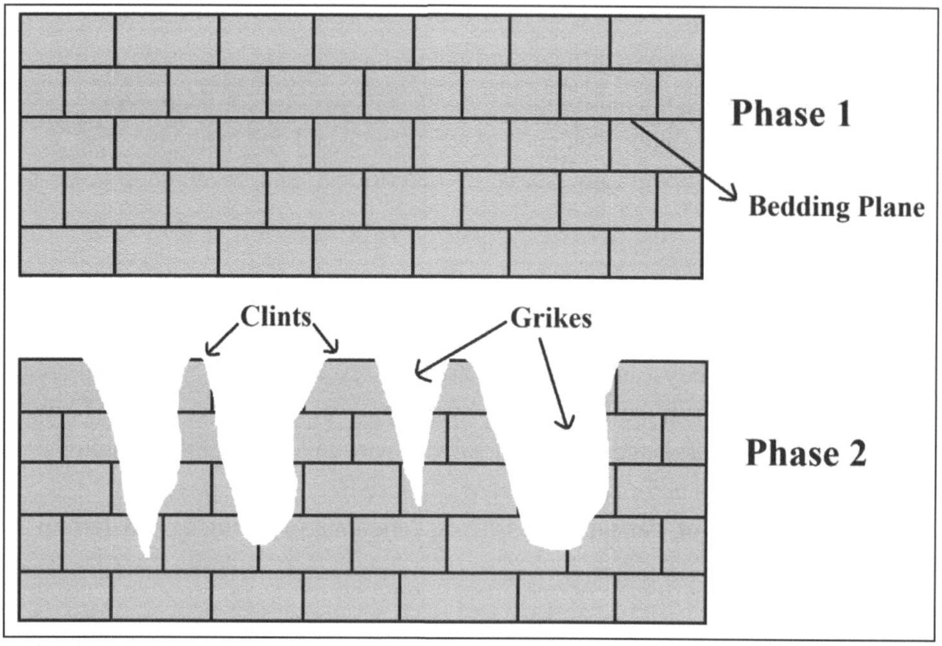

Figure 36.2: limestone pavements are grikes and clints.

Figure 36.1: Image illustrated a limestone pavement (lapies), showcasing the rugged terrain with sharp ridges, deep grooves, and typical features like grikes and clints. The rocky surface is interspersed with small patches of moss and sparse vegetation, reflecting the natural formation of a weathered limestone landscape.

3. **Lapies / limestone pavement:**Lapies, which are also referred to as lapiaz, are surface characteristics that are on a smaller scale and consist of the terrain having deep grooves and sharp ridges. This gives the area a harsh aspect. Along joints, cracks, and bedding planes, these characteristics are produced as a result of the chemical weathering of limestone. A sequence of clefts and ridges are created as a result of the dissolution of limestone by rainwater that is enriched with carbon dioxide. This causes the joints to get deeper. For example, the depth of a lapie may range anywhere from a few centimeters to several meters across. They are often found on rock surfaces that are exposed. Grikes and clints are the two basic components that make up limestone pavements.
4. **Sinkholes (Dolines):** Dolines, another name for sinkholes, are holes or depressions in the earth created when bedrock dissolves at the surface or when the top layer collapses into underlying cavities. The width and depth of a sinkhole may vary from a few meters to several hundred meters. They often act as channels for surface water to submerge underneath.

Types of Sinkholes:

a. **Solution Sinkholes:** This formation is often circular and bowl-shaped, and it is formed by the slow dissolving of the bedrock surface.
b. **Collapse Sinkholes:** Arise as a consequence of the collapse of a cave or hollow located underground.
c. **Subsidence Sinkholes:** Formation that occurs as a result of the slow settling of the earth, which is caused by the passage of loose material downward into cavities below.

5. **Uvalas:** Large depressions known as uvalas are generated when multiple sinkholes those are next to one another combine into a single depression. An uvala is formed when several dolines grow and combine over time, resulting in the formation of a depression that is either irregularly shaped or extended. They are able to cover regions that are as long as several kilometers. It is common for uvalas to have complicated floor topography, including traces of the sinkholes that were there originally.
6. **Poljes:** The term "polje" refers to enormous depressions that are located in karst areas and have a level floor. These depressions have steep sides and may be several kilometers long. Poljes are created as a result of the interaction of three processes: dissolution, subsidence, and tectonic activity. There is also the possibility that the collapse of subsurface caves might have a role in their creation. When it comes to karst landforms, the greatest depressions are called poljes. In many cases, they have a level floor that is covered with sediments, and they may feature lakes that are either

temporary or permanent. Poljes are often drained via swallow holes or drainage systems that are buried underground.

Figure 36.3: Figure illustration of a polje, depicting a large, flat-floored depression typical of karst regions, with steep walls and a mixture of grassy and water-covered areas.

7. **Karst Valleys:** Karst valleys are lengthy depressions or valleys that are formed in limestone areas as a result of the erosive action of flowing water mixed with the disintegration of the bedrock. The surface streams that run through karst valleys may be present for a portion of their course, but they may vanish beneath at some locations. The slopes of the valleys are often steep, and their profiles are not uniform.

Types of Karst valleys

a. **Blind Valleys:** The abrupt termination of these valleys often results in the formation of a sinkhole, which is a place where the surface stream vanishes underground, forming a blind valley.
b. **Dry Valleys:** The redirection of water underground has resulted in the drying out of valleys that were formerly inhabitable by a surface stream but have now become dry.

8. **Caves (Caverns):** The breakdown of soluble rocks along bedding planes, joints, and fractures is the process that brings about the formation of caves, which are naturally

occurring subterranean chambers. Caves are formed when groundwater percolates through joints and bedding planes, causing the rock to dissolve and therefore generating tunnels and chambers that are associated with one another. It's possible that caves possess an intricate network of tunnels and passageways. Stalactites, stalagmites, and columns that were created by the deposition of calcium carbonate are examples of secondary characteristics that are often seen in these formations.

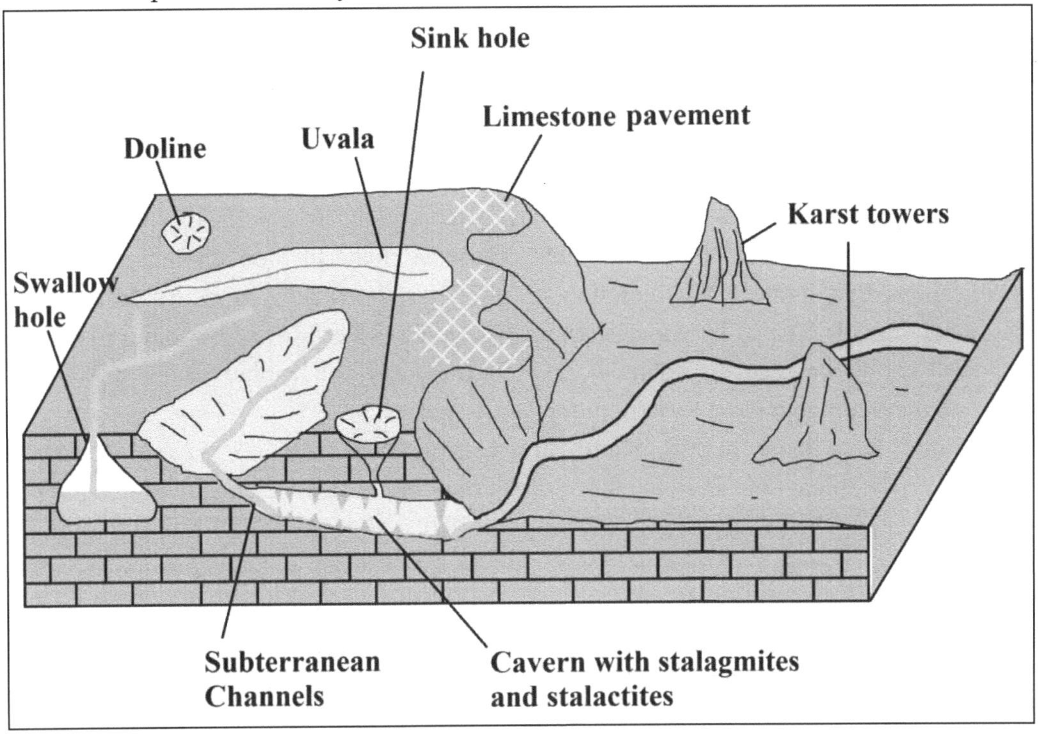

Figure 36.4: Karst Topography

9. **Stalactites, Stalagmites, and Columns (Speleothems):** The deposition of dissolved calcium carbonate is responsible for the formation of these secondary erosional structures that may be seen in the interior of caves. These formations are the result of calcium carbonate that is left behind by water that drops from the roof of a cave. This calcium carbonate accumulates over time to generate these formations. Caverns are made more beautiful by the presence of these formations, which are evidence of active geological processes.

a. **Stalactites:** Formations that resemble icicles that are suspended from the roof of a cave and are created by water pouring down.

b. **Stalagmites:** Shaped like cones, these formations are found just below stalactites and rise up from the cave floor.

c. **Columns:** When stalactites and stalagmites collide and fuse together, this formation is created.

11. **Galleries:** Galleries are passageways or tunnels that link various cave system components. They may be level or moderately sloping. In a karst terrain, they act as connecting passageways between bigger caves, chambers, and other subterranean structures. As groundwater passes through bedding planes, joints, and other weak spots in soluble rocks like limestone, galleries are created. These pathways enlarge over time as the rock is dissolved by groundwater, forming interconnecting tunnels that connect various cave system components.Changes in water table levels, water flow direction, and differences in rock structure may all have an impact on the creation of galleries. In contrast to steep shafts or vertical corridors, galleries are often level or gently sloping. They may range widely in size, from vast sidewalks several meters wide to tiny tunnels. Galleries may include characteristics like scallops that indicate the direction of previous water flow, and they often have smooth walls created by water erosion. Examples:Galleries may be found in several well-known karst cave systems worldwide, including the Jenolan Caves in Australia and the Mammoth Cave System in the United States.
12. **Shaft:** In a karst terrain, shafts are passageways that link several levels of subterranean caverns and chambers. They may be vertical or steeply sloping. They are created when water flows vertically through soluble rock (like limestone) joints, fractures, or cracks. Over time, the tunnel becomes wider as the water moves lower and dissolves the rock. Shafts range in size from massive vertical tunnels to tiny cracks. Shafts might have active water flow (creating subterranean waterfalls) or be dry. Often join horizontal galleries to form a labyrinth of caverns with several levels. Examples: The deep shafts in the Cheve Cave (Mexico) and the "Great Pit" in the Gouffre Berger Cave (France).
13. **Hums or Pepino Hills:** In karst environments, hums—also called Pepino hills—are tiny, remote, residual hills with a dome form that are located on the bottom of enormous depressions like poljes. They are created by the erosion and dissolution of the nearby limestone bedrock, which leaves behind smaller, more robust rock blocks that resemble hills. The isolation of these hums is exacerbated over time by ongoing erosion of the weaker rock, which gives them their characteristic look. Typically, hums have a conical or spherical form. Their heights vary from a few meters to many tens of meters. They are found sporadically on polje floors and resemble solitary hills or groups.
14. **Underground Streams and River Systems:** Surface water vanishes via swallow holes and travels underground, creating underground streams and rivers, which are pathways of flowing water inside a cave system. In order to create subsurface

channels, streams enlarge joints and fractures as they pursue the route of least resistance through the soluble rock. Large passages and chambers may be formed by underground rivers. Eventually, they may continue to flow underground or resurface at the surface via springs.

15. **Stylolite:** A stylolite is a zigzag, uneven, suture-like border that is often seen in sedimentary rocks, such as limestones. Where two neighboring beds interlock, it looks like a serrated line. When minerals dissolve under high pressure along bedding planes or joints, a process known as pressure dissolution, stylolites are created. The distinctive look of the stylolite is caused by the accumulation of less soluble material along its length, such as clay, iron oxides, or organic materials.

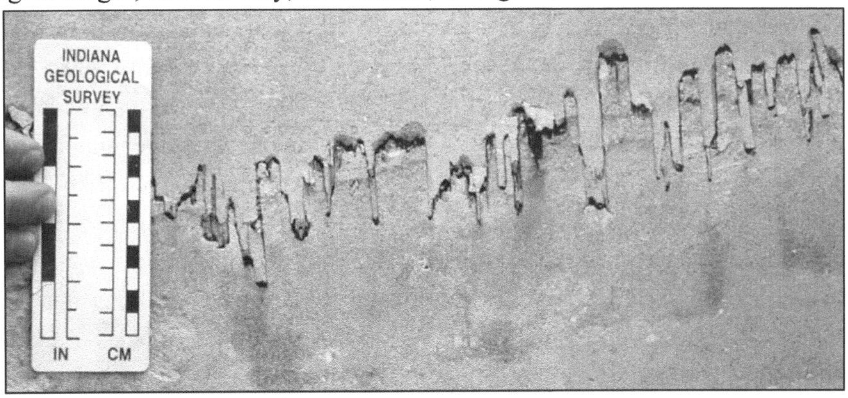

Figure 36.5: Stylolite is an irregular, suture-like, zig-zag boundary in limestone. (Reference: Wikipedia, *the Free Encyclopedia.*)

It appears as a line with interpenetrating peaks and troughs that is wavy or jagged. Insoluble leftovers like clay, quartz, or organic debris may be present on the stylolite surface. It often serves as a reminder of previous pressure and stress levels in the rock strata. Stylolites provide insight into the compaction and stress environment during rock formation by acting as markers of pressure and diagenetic history.

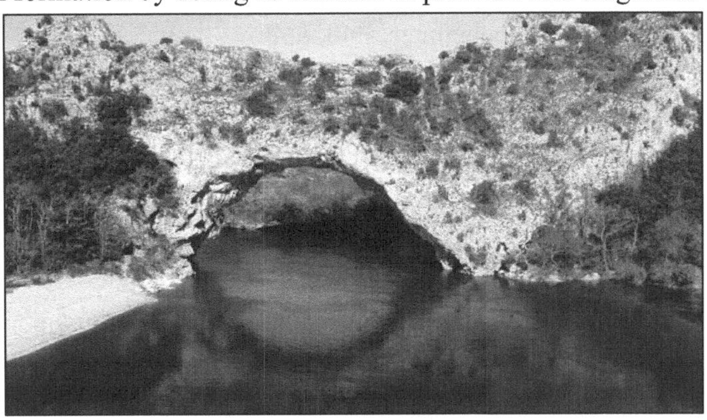

Figure 36.6: Natural Bridges and Arches

16. **Natural Bridges and Arches:** Cave roofs that have not yet fallen have left behind natural bridges and arches such as natural bridges and arches. The formation of these formations occurs when the ceiling of a cave in part collapses, leaving behind a section of rock that spans the open area after the collapse. In mature karst environments, these characteristics are often seen. They are able to change in size and form depending on the amount of erosion that has occurred.
17. **Swallo Holes:** Swallow holes are tiny apertures or depressions in the earth that enable surface water from rivers or streams to vanish beneath and join a cave system or network of subterranean channels. Limestone and other soluble rocks are prevalent in the areas where it is created. As surface streams come into contact with joints, cracks, or other weak areas in the bedrock, the water begins to dissolve the rock, which ultimately results in the opening becoming steadily larger. When these little apertures are allowed to grow over time, they eventually create swallow holes, which are responsible for channeling surface water into subterranean channels. It is often discovered in regions that are characterized by a large number of sinkholes or along the path of sinking streams. There is a wide range of sizes for them, from tiny holes to enormous apertures that are capable of consuming whole streams. It is possible for swallow holes to go into caverns and subsurface drainage systems that are rather sophisticated. Due to the fact that they serve as entrance places for water into underground channels, they contribute to the formation of cave systems that are located underground. Swallow holes are often indicative of the beginning of a karst drainage system, which is characterized by the disappearance of surface water and the subsequent movement of water via subterranean streams.

Depositional Features

1. **Geode**: A geode is a sort of rock formation that is hollow and spherical, and it typically has an exterior layer of chalcedony or another type of mineral, and it is lined with crystals that face inward. The formation of this phenomenon occurs when mineral-rich groundwater penetrates into hollow crevices within sedimentary or igneous rocks. Crystallization of minerals such as quartz or amethyst occurs inside the cavity over the course of time. The cavity of the geode is lined with crystals that have been formed exceptionally well and protrude inward, giving the geode the appearance of a brilliant jewel.
2. **Sinter**: Sinter is a deposit that may be composed of either silica or calcium carbonate. It is generated when precipitation from hot springs, geysers, or groundwater makes its way into the earth. The precipitation of dissolved minerals results in the formation of a porous, stratified deposit. This occurs when mineral-rich fluids, whether hot or cold, continue to cool or evaporate.

Types of Sinter:

a. **Siliceous Sinter**: Characterized by the presence of silica (SiO_2), this is often found in geysers and hot springs.
b. **Calcareous Sinter (Travertine)**: This mineral is formed from calcium carbonate ($CaCO_3$), typically found in regions with limestone bedrock and linked with hot springs.
c. **Kanker**: Kanker is a colloquial name that is often used in South Asia to refer to concretions, which are deposits of hard, cemented gravel that are found in alluvial environments. These are produced when loose gravel and alluvium are treated to cementation and compaction, which is often accomplished by the use of calcium carbonate, iron oxide, or silica respectively. These minerals are typically found in soils or sedimentary layers in arid or semi-arid climates, where changes in groundwater contribute to the precipitation of these minerals.

Chapter 37
Lakes

Lakes are bodies of water that are found on the surface of the Earth and are located in depressions. Lakes may be as tiny as ponds or as large as enormous basins that contain either freshwater or saltwater. Lakes are considered to be ephemeral features in a landscape from a geological point of view because they are sensitive to the processes of sedimentation, erosion, and fluctuations in climate. The study of lakes, or limnology, is vital for understanding the dynamics of sediment deposition, biological niches, and hydrological cycles. In this chapter, we will investigate numerous kinds of lakes, their origins, and their geological relevance.

Definition and Basic Characteristics

- The term "lake" refers to a body of water that is generally vast in size, has a definite border, and is located in a land depression. It is possible for lakes to contain either freshwater or saltwater, and they may vary widely in size, depth, and form.
- The classification of lakes is determined by a number of criteria, including their places of origin, hydrological properties, and salinity.

Geological Classification of Lakes

The primary factor that geologists use to categorize lakes is their origin, which is a reflection of the processes that generated the depressions that they inhabit. There are many kinds of lakes that may be classified according to their geological characteristics. These include Tectonic Lakes, Volcanic Lakes, Glacial Lakes, Fluvial Lakes, Karst Lakes, Aeolian Lakes, Solution Lakes, Landslide and Barrier Lakes. Let's go further into each category by providing geological explanations and actual instances.

1. **Tectonic Lakes**: It is the motions of the Earth's crust, such as faulting, warping, or rifting, that are responsible for the formation of tectonic lakes. Divergent plate borders, fault zones, and subsidence areas are all potential locations where these structural depressions might emerge.

Geological Processes:

a. Faulting: Rough valleys or grabens are formed, which have the potential to get filled with water and produce vast linear lakes.
b. Warping: Causes basins that have the potential to collect both surface runoff and groundwater.
c. Subsidence: Happens as a result of crustal sagging or tectonic downwarping.

Examples:

a. Lake Baikal (Russia): Located in a rift valley, this tectonic lake is the deepest and oldest of its kind in the world.
b. Lake Tanganyika (Africa):A lake that was produced by tectonic processes along the East African Rift and takes the shape of a rift valley.

   Geological Significance: Tectonic lakes are significant for the study of geological history, sediment deposition, and tectonic development since they often have a large depth and are quite old.

2. **Volcanic Lakes**: The craters or calderas of dormant or extinct volcanoes are the habitats for the formation of volcanic lakes. They are also capable of developing in depressions that are caused by impediments in the flow of lava.

Geological Processes:

a. Caldera Collapse: The empty magma chamber falls after a volcanic eruption, forming a sizable depression that may eventually fill with water (such as Crater Lake).
b. Lava Flow Damming: Lake Creation occurs when lava flows block river valleys, forming a barrier.

Examples:

a. Crater Lake (USA): A traditional caldera lake created when Mount Mazama collapsed.
b. Lake Toba (Indonesia): a caldera lake that was created 74,000 years ago by a huge volcanic explosion.

Geological Significance: In addition to having distinctive chemical compositions because of dissolved volcanic gases, volcanic lakes may provide information about the dynamics of magma chambers and historical volcanic activity.

3. **Glacial Lakes**: The motion of glaciers creates glacial lakes. They may be found in both current alpine regions and locations that were formerly covered by ice sheets.

Geological Processes:

a. Glacial Erosion: Causes fjords, valleys, or cirques to form, which are subsequently filled with meltwater.

b. Moraine Damming: Meltwater may be trapped behind glacial till (moraines) that have accumulated, acting as a dam.

Examples:

a. Great Lakes (USA and Canada): Formed during the last Ice Age by moraine accumulation and glacial scouring.

b. Lake Geneva (Switzerland): A glacier lake created by the Rhône Glacier's activity.

Geological Significance: Glacial lakes may provide information about former glacial extents and paleoclimates, making them sensitive markers of climate change.

3. **Fluvial Lakes**: Oxbow lakes, floodplain lakes, and naturally occurring levee lakes are examples of fluvial lakes that are created by river activity.

Geological Processes:

a. Meander Cutoff: An isolated body of water called an oxbow lake is created when a river meander is cut off.

b. Floodplain Deposition: Water may fill up depressions created by sediment deposition in floodplain zones.

Examples:

a. Oxbow Lakes along the Mississippi River (USA).

b. Loktak Lake (India): A floodplain lake formed by the deposition of sediment in the Manipur River basin.

Geological Significance: Fluvial lakes are transient landforms that provide light on channel migration, sediment transport, and fluvial dynamics.

4. **Karst Lakes**: In areas of limestone or dolomite, when sinkholes or depressions are created by groundwater dissolution, karst lakes arise.

Geological Processes:

a. Solution Weathering: Acidic water dissolves soluble rocks, creating dolines and sinkholes that might eventually fill with water.

b. Collapse of Caverns: Depressions that function as basins may be formed by underground cavern collapses.

Examples:

a. Lake Cerknica (Slovenia): A karst lake that fills up in the rainy season and goes dry in the dry season.

b. Pichola Lake (Udaipur, India): A karst lake formed due to solution processes in the underlying marble.

Geological Significance: The intricate interactions between surface water dynamics, rock breakdown, and groundwater movement are reflected in karst lakes.

5. **Aeolian Lakes**: Wind erosion in arid areas produces depressions that may hold water, resulting in the formation of aeolian lakes.

Geological Processes:

a. Wind Erosion: When loose sediment is removed by wind, depressions known as deflation hollows or pans are formed.

b. Dune Damming: Basins may be formed when valleys are blocked by wind-blown sand dunes.

Examples:

a. Chott el Jerid (Tunisia): A shallow aeolian salt lake in the Sahara Desert.

b. Sambhar Lake (India): A saline lake formed by aeolian and fluvial processes.

Geological Significance: Aeolian lakes serve as markers of current and historical wind patterns and climatic conditions.

6. **Solution Lakes**: When chemical weathering processes produce depressions in areas with soluble bedrock (such as gypsum or limestone), solution lakes are formed.

Geological Processes: Acidic groundwater causes soluble minerals to dissolve, causing cracks and joints to enlarge forming depressions that might hold water.

Examples:

a. Dead Sea (Israel-Jordan): A solution lake created by salt dissolving and its location inside a tectonic basin is the Dead Sea (Israel-Jordan).

b. Salton Sea (USA): A solution lake created by geological and chemical processes in a depression.

Geological Significance: Solution lakes often have high salinity and unique ecological characteristics due to their chemistry.

7. **Landslide and Barrier Lakes**: Landslide or barrier lakes are formed when landslides, debris flows, or volcanic eruptions block valleys, creating impounded water bodies.

Geological Processes:

a. Landslide Damming: A temporary or permanent lake is created when a mass of rock or dirt blocks the flow of a river.

b. Volcanic Lava Flows: May create lakes by obstructing river routes.

Examples:

a. Lake Sarez (Tajikistan): Formed by a massive landslide in 1911.

b. Lake Batur (Indonesia): Created by volcanic lava flows.

Geological Significance: These lakes are susceptible to seismic activity and are vulnerable to devastating outburst floods.

8. **Anthropogenic (Man-made) Lakes**: Lakes that are man-made are produced by damming rivers, mining, or other building operations.

Geological Processes:

a. River Damming: For water supply, agriculture, or hydroelectricity.

b. Mining Pits: Filling up mining pits or abandoned quarries.

Examples:

a. Lake Mead (USA): Formed by the Hoover Dam on the Colorado River.

b. Gobind Sagar Lake (India): Formed by the Bhakra Nangal Dam on the Sutlej River.

Geological Significance: Man-made lakes are important for ecological preservation, energy generation, and water resource management.

Changing geological and climatic circumstances cause lakes to change as dynamic landforms. An understanding of Earth's geological processes and environmental changes may be gained by researching their origin, sedimentation patterns, and hydrology.

Chapter 38
Oceans and their landforms

Introduction to Ocean and Coastal Geography

Geological, hydrological, and atmospheric processes are responsible for shaping the oceans and coastlines, which are dynamic ecosystems that are home to a broad diversity of physical characteristics. It is essential to have a solid understanding of these characteristics in order to have a complete comprehension of the marine and coastal systems of the Earth, as well as their ecological relevance and influence on human activities. In this chapter, we will discuss the many physical characteristics of seas and beaches, as well as their creation and the relevance of these characteristics.

Physical Features of Oceans

1. **Continental Margins:** The submerged outside borders of a continent, which are located at the point where the land meets the water, are known as continental margins. They are the transitional layer that exists between the oceanic crust and the continental crust. There are three primary components that may be extracted from continental margins:

a. **Continental Shelf:** The continental shelf is the portion of the margin that is mildly sloping and rather shallow. It stretches from the coastline to the point where the slope starts to steepen, which is referred to as the continental slope. Because of the relatively shallow waters, it is often covered with silt and has the potential to provide a habitat for a diverse array of marine species. For the most part, the depths may range anywhere from 0 to around 200 meters (660 feet), but the breadth can change. Oil, gas, and fisheries are just some of the natural resources that may be found in abundance on continental shelves.

b. **Continental Slope:** The edge of the continental shelf is denoted by the continental slope, which is also the point at which the ocean floor begins to drop in a steep manner. It makes a rapid descent from the shelf break, which is the point at which the slope begins, to the deeper ocean basin. The slope is steeper than the shelf, and it frequently descends to depths that are several kilometers deep. Submarine canyons, which are frequently carved by turbidity currents (underwater landslides), are often discovered along the slope.

c. **Continental Rise:** The more gradual incline known as the continental rise is located beyond the continental slope. Significant sedimentary deposits have been formed by the accumulation of sediments that have slid down from the shelf and slope. Usually, the rise indicates the boundary between the deep ocean floor (the abyssal plain) and the continental crust.

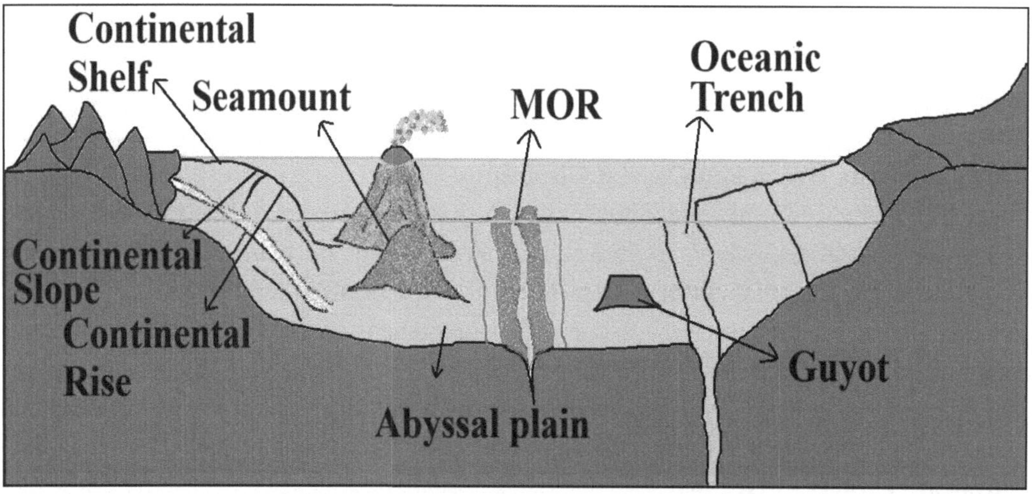

Figure 38.1: topographic features of ocean basins

3. **Deep Sea Trenches:** Long, narrow depressions in the ocean floor that frequently reach depths of over 6,000 meters are known as deep sea trenches. They are connected to tectonic plate borders and are the deepest regions of the ocean.

A. **Formation**

a. **Subduction Zones**

- Process: When two tectonic plates collide, one is driven beneath the other into the mantle, creating deep sea trenches. This process is known as subduction.

- Mechanism: A trench is formed when the less dense continental or oceanic plate descends beneath the denser oceanic plate.

b. Tectonic Activity

- Earthquakes: There is a high frequency of large earthquakes in subduction zones, which are regions that are highly seismic.
- Volcanism: Connected to volcanic arcs that originated on the plate that is overriding it.

C. Examples of Deep Sea Trenches

a. Mariana Trench

- Location: Western Pacific Ocean.
- Depth: The Challenger Deep, the lowest known point on Earth, is about 11,034 meters deep.
- Significance: Extreme pressure conditions and distinct ecosystems can be found there.

b. Peru-Chile Trench (Atacama Trench)

- Location: Off South America's western coast.
- Depth: Reaches approximately 8,065 meters below the surface.
- Significance: Connected to significant earthquakes and the Andean volcanic arc.

c. Java Trench (Sunda Trench)

- Location: To the south of Indonesia, in the Indian Ocean
- Depth: Can be found at depths of approximately 7,725 meters.
- Significance: The epicenter of significant seismic activity, including the earthquake and tsunami that occurred in the Indian Ocean in 2004.

D. Geological and Biological Significance

a. Geological Processes

- Plate Recycling: The process of recycling oceanic crust into the mantle takes place at trenches.
- Seismic Activity: They are important places to study earthquake dynamics and processes related to subduction because of their specific locations.

b. Biological Diversity

- Extreme Environments: Habitate organisms that are one of a kind and have adapted to conditions of high pressure, low temperature, and low light.
- Chemosynthesis**:** There are a great number of deep-sea species that obtain their energy from chemosynthetic bacteria, particularly in the vicinity of hydrothermal vents.

A. Mid-Oceanic Ridges: Mid-oceanic ridges are mountain ranges that are located underwater and are formed as a result of tectonic activity at the boundaries of divergent plates. A central rift valley and extensive volcanic activity are two of the distinguishing features of these regions.

Formation

a. **Seafloor Spreading**

- **Process:** The formation of mid-ocean ridges occurs at divergent plate boundaries, which are the points at which two tectonic plates move apart.
- **Mechanism:** Magma rises from the mantle to fill the void, and as it cools and solidifies, it brings about the formation of new oceanic crust.

b. **Volcanism**

B. **Underwater Volcanos:** Volcanic activity that occurs continuously along the ridge results in the formation of new crust and contributes to the elevation of the ridge.

Examples of Mid-Oceanic Ridges

a. **Mid-Atlantic Ridge**

- Location: Is a body of water that extends from the Arctic Ocean to the Southern Ocean and runs through the middle of the Atlantic Ocean.
- Characteristics: It is characterized by frequent volcanic activity and a rift valley located at its peak.
- Significance: Plays a significant part in the theory of plate tectonics and the spreading of the seafloor.

b. **East Pacific Rise**

- Location: Stretches from the Gulf of California all the way up to the Pacific-Antarctic Ridge in terms of its location.
- Characteristics: Ridge that is rapidly spreading and contains a high level of volcanic activity and hydrothermal vents.
- Significance: It is essential for the study of hydrothermal ecosystems and the dynamics of fast-spreading ridges.

c. **Indian Ocean Ridge**

- Location: Consists of the Central Indian Ridge, Southeast Indian Ridge, and Southwest Indian Ridge.
- Characteristics: A network of interconnected ridges with different rates of spread.
- Significance: Impacts the Indian Ocean basin's geological and tectonic development.

C. **Geological and Biological Significance**

a. **Geological Processes**

- Crust Formation: New oceanic crust is constantly being formed on mid-oceanic ridges.
- Heat Flow: Important geothermal heat sources that affect global and oceanic temperature trends.

b. **Biological Diversity**

- Hydrothermal Vents: Support distinct chemosynthesis-dependent ecosystems that include specialized bacteria, clams, and tube worms.
- Endemic Species: There are a lot of species that are unique to mid-oceanic ridges.

F. Abyssal Plains: The flat, almost featureless areas of the deep ocean floor are known as abyssal plains. They are some of the planet's most vast and least diverse ecosystems.

Formation

a. Sediment Deposition

- **Process:** Fine-grained sediments, such as clay, silt, and organic matter, accumulate from the water column to form abyssal plains.
- **Sources:** Pelagic (open ocean) sources, terrestrial runoff, and biogenic material from marine organisms are the sources of sediments.

b. Tectonic Activity

- **Plate Movements:** Usually found between continental margins and mid-oceanic ridges, abyssal plains cover comparatively dormant regions of tectonic plates.

G. Examples of Abyssal Plains

a. Blake Plateau

- **Location:** Off the southeast coast of the United States is the Western Atlantic Ocean.
- **Characteristics:** Renowned for having thick sediment deposits and being shallower than other abyssal plains.

b. Sohm Abyssal Plain

- **Location:** South of Newfoundland, in the North Atlantic Ocean.
- **Characteristics:** Thick layers of fine sediments cover one of the biggest and flattest abyssal plains.

c. Peru Basin

- **Location:** Southeastern Pacific Ocean.
- **Characteristics:** Includes deep-sea sediment layers and vast fields of manganese nodules.

H. Geological and Biological Significance

a. Geological Processes

- **Sediment Accumulation:** Global sediment cycles are influenced by the enormous volumes of sediments that are stored in abyssal plains.
- **Tectonic Stability:** A long-term record of sedimentation and oceanic conditions can be found in these areas, which are usually tectonically stable.

b. Biological Diversity

- **Deep-Sea Ecosystems:** A wide range of creatures, including benthic (bottom-dwelling) species like polychaete worms, sea cucumbers, and isopods, can be found in abyssal plains.

- **Nutrient Cycling:** Abyssal plain sediments' organic matter is essential for carbon sequestration and deep-sea nutrient cycling.

4. **Seamounts:** Underwater Mountains created by volcanic activity are known as seamounts. Though they do not reach the surface, these mountains rise from the ocean floor. They are found in all of the world's oceans and usually have a conical or peaked shape. Underwater volcanic activity results from magma pushing through the Earth's crust to form seamounts. A mountain is formed over time by the accumulation of layers of hardened lava from these eruptions.
5. **Guyots**: There is a type of seamount known as a guyot that has a flat top.Wave action, which occurred when the seamount was once located above the ocean's surface or in close proximity to it, typically causes erosion, leading to the flat top. First, guyots form similarly to regular seamounts. As a result of periods of volcanic activity, some of them rise to the surface of the water or come very close to it, evolving into islands. Tectonic movements cause the ocean floor to continue to spread and sink, which causes the island to submerge over time, leaving behind a flat-topped seamount beneath the surface. Wave erosion gradually flattens the summit over time.

Types of Continental Margins:

There are two major types of continental margins, based on tectonic activity:

1. **Active Margins** (Pacific-type margins): It is typically found at subduction zones, where two plates are being forced beneath one another, along tectonic plate boundaries. Earthquakes, volcanic activity, and the development of deep ocean trenches are characteristics of these margins. Example: The western coasts of North and South America.
2. **Passive Margins** (Atlantic-type margins):It is not found close to a plate boundary; rather, it is found along tectonic plate boundaries that are not active. Broad continental shelves, slopes, and rises define these margins, which are characterized by minimal tectonic activity. Example: The eastern coasts of North and South America, as well as the coastlines of Africa and Europe.

Ocean Currents

Large-scale flows of seawater that travel through the oceans of the world are referred to as ocean currents. The rotation of the Earth, wind, temperature, and differences in salinity are some of the factors that contribute to their movement.

1. **Surface Currents**: Currents that are horizontal and driven by the wind that are found in the uppermost 400 meters of the ocean. Examples: Gulf Stream, Kuroshio Current.

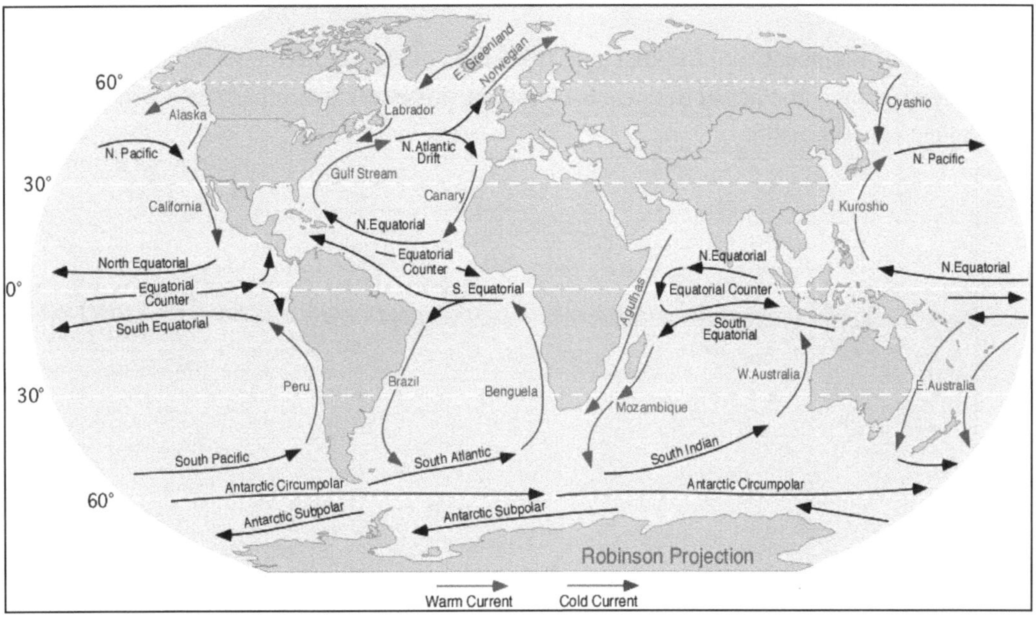

Figure 38.2: ocean Surface Currents (Reference: Wikipedia, *the Free Encyclopedia.*)

2. **Deep Water Currents:** Thermohaline circulation refers to the movement of water in both a vertical and horizontal direction, which is driven by differences in density that are caused by variations in temperature and salinity. Examples: Antarctic Bottom Water, North Atlantic Deep Water.

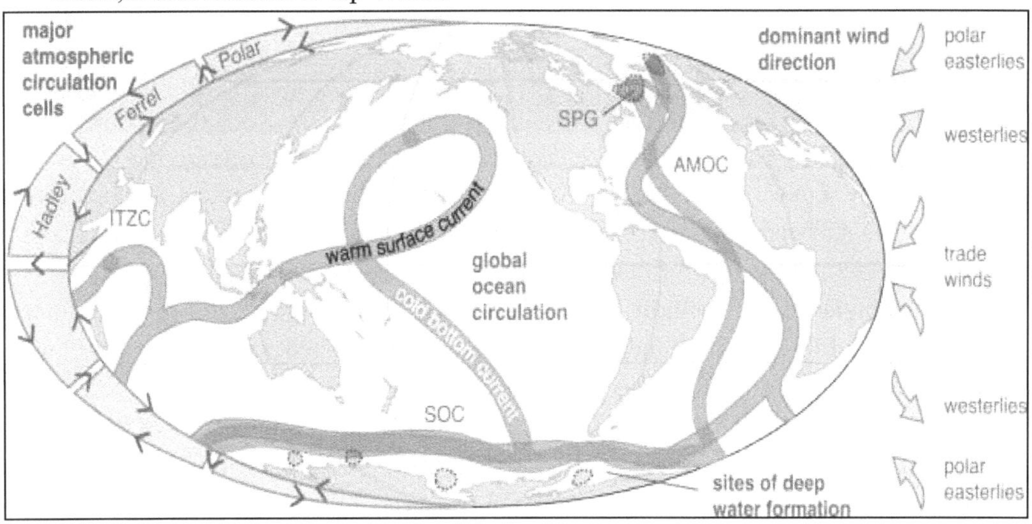

Figure 38.3: Deep Water Currents of ocean (Reference: Wikipedia, *the Free Encyclopedia.*)

Oceanic Zones

The depth of the ocean and the distance from the coast provide the basis for the division of the ocean into distinct zones, each of which possesses a unique set of physical and biological characteristics.

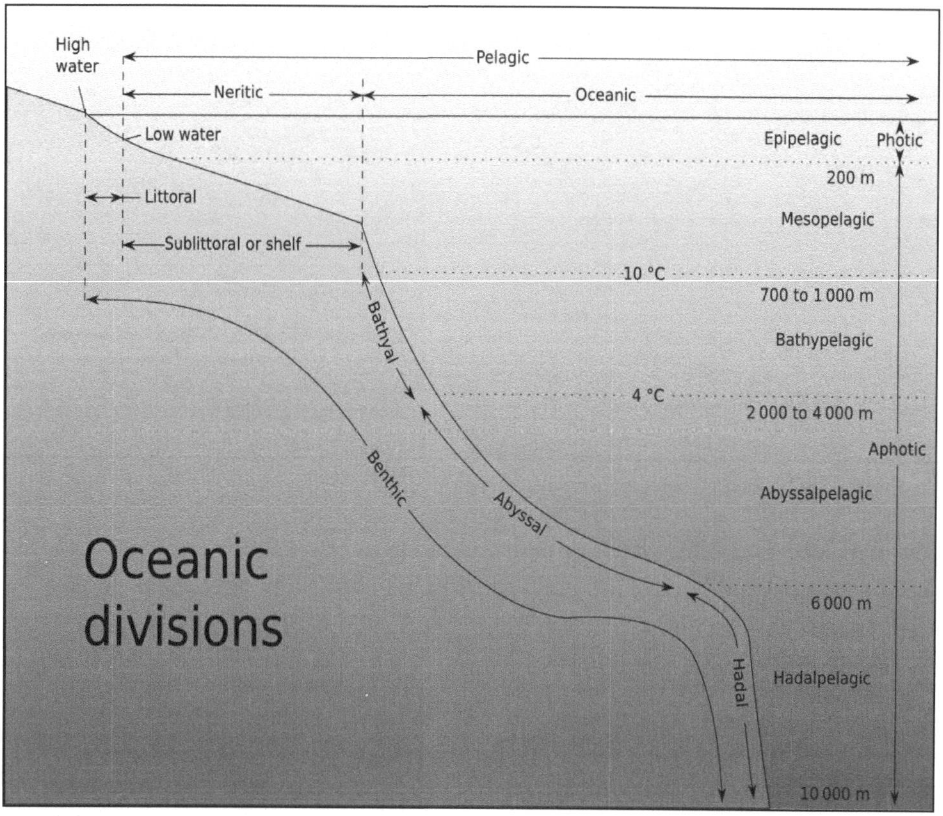

Figure 38.4: Oceanic Zones (Reference: Wikipedia, *the Free Encyclopedia.*)

1. **Epipelagic Zone (Sunlit Zone):** A depth range of 0 to 200 meters is available. It receives a sufficient amount of sunlight, which is necessary for photosynthesis, and it is home to a wide variety of marine life.
2. **Mesopelagic Zone (Twilight Zone)**: There is a depth range of 200 to 1,000 meters to it. In the Mesopelagic Zone, light penetration is restricted, temperatures are lower, and marine species have adapted to their environment.
3. **Bathypelagic Zone (Midnight Zone)**: The range of its depth is from 1,000 to 4,000 meters. In this region, there is no sunlight, the pressure is high, and there are bioluminescent organisms.
4. **Abyssopelagic Zone (Abyss)**:From 4,000 to 6,000 meters is the range of its depth. This zone has temperatures that are close to freezing, extremely high pressure, and very little life.

5. **Hadalpelagic Zone (Trenches):** To the bottom of the ocean, it extends to a depth of 6,000 meters. Extreme conditions and very specific organisms are found in the Hadalpelagic Zone.

Physical Features of Coasts

Coastal Landforms: The interactions between land, sea, and atmospheric processes are responsible for the formation of coastal landforms, which results in environments which are both diverse and dynamic.

1. **Beaches**
- Description: Beaches are defined as the accumulation of sand, gravel, or pebbles along the boundary of the shoreline.
- Formation: These sediments are the result of waves, tides, and currents depositing sediments on the ocean floor.
- Types: Sandy beaches, shingle beaches.

2. **Cliffs and Wave-Cut Platforms**
- Description: The steep rock faces that run along the coast and the flat areas that are at their base are referred to as cliffs and wave-cut platforms.
- Formation: erosion caused by waves, which results in the retreat of cliffs and the formation of platforms.
- Examples: White Cliffs of Dover (UK), Cliffs of Moher (Ireland).

3. **Dunes**
- Description: Sand ridges or mounds that are formed as a result of wind deposition.
- Formation: The sand that is carried and deposited inland from beaches is caused by wind.
- Significance: Inhibit the erosion of coastal areas by acting as natural barriers.

4. **Estuaries**
- Description: Inlets along the coast where freshwater from rivers combined with saltwater from the ocean.
- Formation: Originating from river valleys that have been submerged or tectonic depressions
- Examples: Chesapeake Bay (USA), Thames Estuary (UK).

5. **Barrier Islands**
- Description: A lagoon or bay serves as a barrier between the long, narrow islands that run parallel to the coast.
- Formation: The wave action, the deposition of sediment, and the rising sea levels all contribute to this result.
- Examples: Outer Banks (USA), Great Barrier Island (New Zealand).

6. **Deltas**

- Description: Formations of land that are formed at the mouths of rivers as a result of the deposition of sediment.
- Formation: When rivers enter bodies of water that are still, sediments that they have carried with them are deposited.
- Types: Arcuate, bird's foot, cuspate.
- Examples: Nile Delta (Egypt), Mississippi Delta (USA).

7. **Lagoons**

- Description: Bodies of water that are relatively shallow and are separated from the ocean by a barrier, such as a sandbar or a coral reef.
- Formation: Produced by the accumulation of sediments or the growth of coral.
- Examples: Venice Lagoon (Italy), Blue Lagoon (Iceland).

B. Coastal Processes

The activities of waves, tides, and currents that are responsible for shaping the coastal landscape are referred to as coastal processes.

1. **Wave Action**

a. Erosion: Hydraulic action, abrasion, and solution are the three processes that cause waves to erode the coastline.

b. Transportation: The movement of sediments along the coast is caused by longshore drift.

c. Deposition: Beaches, spits, and bars are all formed by the sediments that are deposited by waves.

2. **Tidal Action**

a. Tidal Currents: Sediments are eroded, transported, and deposited based on the currents that are generated by tides.

b. Tidal Flats: Areas of sediment deposition in intertidal zones, which are frequently abundant in a variety of organisms.

3. **Currents**

a. Longshore Currents: Sediments are being moved by flow that is parallel to the coast.

b. Rip Currents: Currents that are strong and narrow, flowing from the shore to the ocean, and are formed by water that is returning to the ocean.

C. Human Impact on Coasts

Coastal environments are significantly impacted by human activities, which frequently result in conditions such as erosion, the loss of habitat, and changes in the dynamics of sediment.

1. **Coastal Development**

- Impact: Habitat destruction and increased erosion are two potential outcomes that can result from the construction of buildings, roads, and other infrastructure.
- Examples: The urbanization of coastlines in cities such as Miami (The United States of America) and Sydney (Australia).

2. **Sea-Level Rise**

- Impact: Climate change causes sea levels to rise, which in turn leads to an increase in coastal erosion, flooding, and the loss of large amounts of land.
- Examples: Threatened coastal areas in the Maldives, Bangladesh.

3. **Pollution**

- Impact: Coastal waters are polluted by runoff from urban areas, industrial areas, and agricultural land, which has a negative impact on marine life and ecosystems.
- Examples: Dead zones in the Gulf of Mexico, plastic pollution in the Pacific Ocean.

4. **Coastal Management**

- Approach: Hard engineering, such as seawalls and groynes, and soft engineering, such as beach nourishment and dune restoration, are both considered to be measures that are taken to protect coastlines.
- Examples: Coastal defenses in the Netherlands, beach replenishment in Florida (USA).

Case Studies of Coastal and Oceanic Features

A. The Great Barrier Reef (Australia)

- Description: Over 2,900 individual reefs make up the world's largest coral reef system, which is situated in the Pacific Ocean.
- Significance: The area is a major tourist destination, a biodiversity hotspot, and it provides a habitat for marine life.
- Threats: Overfishing, pollution, climate change, and coral bleaching are all growing concerns.

B. The Bay of Fundy (Canada)

- Description: The tidal range of this area is known to be the highest in the world, reaching up to 16 meters.
- Significance: Tidal phenomena that are one of a kind produce abundant ecosystems and opportunities for the generation of renewable tidal energy.
- Challenges: The management of sedimentation and erosion alongshore.

C. The Mississippi Delta (USA)

- Description: Deposition of sediments from the Mississippi River has resulted in the formation of a delta known as the bird's foot delta.
- Significance: Important for the preservation of biodiversity and fisheries, as well as for acting as a barrier against hurricanes.

- Threats: Levees and canals are examples of human-made modifications that contribute to subsidence and rising sea levels.

D. The Maldives

- Description: An archipelago in the Indian Ocean consisting of over one thousand coral islands and consisting of twenty-six atolls.
- Significance: Admired for the richness of its marine life and the tourism industry
- Challenges: vulnerable to the effects of rising sea levels, coral bleaching, and erosion along the coast.

Conservation and Sustainable Management of Ocean and Coastal Environments

A. Marine Protected Areas (MPAs)

- Purpose: Protect endangered species, preserve marine biodiversity, and ensure the continued viability of fisheries.
- Examples: Great Barrier Reef Marine Park (Australia), Papahānaumokuākea Marine National Monument (USA).

B. Coastal Zone Management

- Approach: Integrated Coastal Zone Management (ICZM) encourages the conservation of coastal resources and the use of those resources in a sustainable manner.
- Strategies: The balancing of development and conservation, the participation of stakeholders, and the implementation of adaptive management.

C. Restoration Projects

- Purpose: Restore habitats that have been degraded along the coast and in the ocean, such as coral reefs, mangroves, and wetland areas.
- Examples: Mangrove restoration in Bangladesh, oyster reef restoration in the Chesapeake Bay (USA).

D. Climate Change Mitigation and Adaptation

- Mitigation: The reduction of emissions of greenhouse gases in order to curb the rise in sea level and global warming.
- Adaptation: Developing infrastructure that is resilient, protecting natural barriers, and planning for the impacts of rising sea levels are all important activities.

The complex interactions that occur between geological, hydrological, and atmospheric processes are responsible for the formation of the geomorphological characteristics of oceans and coasts. Deep ocean basins, mid-ocean ridges, coastal cliffs, and barrier islands are all examples of features that are essential to the ecosystems and human societies that exist on Earth. For effective management and conservation in the face of increasing human impact and climate change, it is essential to have a solid understanding of these features and the processes that led to their formation within the environment. Through the study and protection of our coastal and oceanic environments, we can ensure that they will continue to be healthy and sustainable for future generations.

Chapter 39
Major geomorphic features of India

Coastal, peninsular, and extra peninsular areas are the major categories that are used to classify the geomorphic characteristics seen in India. Over the course of millions of years, tectonic action, weathering, erosion, and deposition have all contributed to the formation of a wide variety of landforms in these locations. A brief summary of the most prominent geomorphic characteristics found in each area is as follows:

1. **Coastal Geomorphic Features of India:** India has a vast coastline that stretches for around 7,516 kilometers and is comprised of a variety of geographical features along the shore. The western coast, which runs along the Arabian Sea, and the eastern coast, which runs along the Bay of Bengal, are the two distinct sections of the coastline. Among the most important coastline characteristics are:

A. Western Coastal Plain

a. Narrow and Steep: As compared to the eastern shore, the western coast is more narrow and steep.
b. Konkan Coast: This part of the north runs from Maharashtra to GOA. There are several streams, estuaries, and little natural harbors that define it.
c. Kannada Coast: Also referred to as the Kanara or Karavali coast, this Karnataka region is well-known for its cliffs and sandy beaches.

d. Malabar Coast: Malabar Coast: This southern region, which crosses Kerala, is home to wetlands, estuaries, and lagoons, often referred to as backwaters. One of the most notable backwater systems in this area is the Vembanad Lake.

e. Coastal Sand Dunes: Sand dunes created by wind activity may be seen on sandy beaches throughout parts of the western coast, particularly in Kerala and Goa.

B. Eastern Coastal Plain

a. Broad and Gently Sloping: In contrast to the western shore, the eastern coast is wider and slopes more gently.

b. Coromandel Coast: Deltas created by rivers like the Krishna, Kaveri, and Godavari define this region, which stretches from Tamil Nadu to southern Andhra Pradesh. There are several estuaries and sandy beaches along this coast.

c. Utkal Coast: The Mahanadi delta marks the northern portion of the eastern coast, which is located in Odisha. The Chilika Lake, Asia's biggest coastal lagoon, is one of the key characteristics.

d. Deltas: Fertile alluvial deltas created by significant rivers like the Ganga, Godavari, Krishna, and Mahanadi are well-known along the eastern coast. The biggest delta in the world and a UNESCO World Heritage Site is the Sundarbans, which was created by the Ganges and Brahmaputra rivers.

C. Coral Reefs and Islands

a. Lakshadweep Islands: These islands, which make up a significant marine environment, are made up of coral reefs, atolls, and tiny islands in the Arabian Sea.

b. Andaman and Nicobar Islands: These volcanic islands are situated in the Bay of Bengal. The only active volcano in India is Barren Island. Around certain islands, coral reefs may also be found.

c. Coral Reefs: Coral reefs may also be found between Tamil Nadu and Sri Lanka, near the Palk Strait and the Gulf of Mannar.

2. Peninsular Geomorphic Features of India: In southern India, the Peninsular Plateau is a large, stable landmass that was mostly composed of old, crystalline rocks. Plateaus, hill ranges, and river basins define its center area, while the Eastern and Western Ghats form its borders. Among the main peninsular characteristics are:

A. Deccan Plateau

a. Extent: A large portion of southern India is covered by this enormous triangular plateau. It stretches from the northern Satpura mountain to the southernmost point of the peninsula.

b. Volcanic Origin: The Deccan Traps, which are strata of volcanic basalt rock created by ancient lava flows, are well-known in the northwest, particularly in the area around Maharashtra.

c. Western Ghats: This range, which runs parallel to the western shore, is also known as the Sahyadris. The Western Ghats are one of the world's hotspots for biodiversity and a UNESCO World Heritage site. Anamudi in Kerala is the highest mountain at 2,695 meters.

d. Eastern Ghats: Parallel to the eastern shore, this range is less continuous and lower in height than the Western Ghats. Rivers such as the Godavari, Krishna, Kaveri, and Mahanadi cut it.

e. Nilgiri Hills: The Nilgiri Hills, which are renowned for their abundant biodiversity, are situated where the Western and Eastern Ghats converge. Doddabetta is the highest summit (2,637 meters).

B. Central Highlands

a. Aravalli Range:TheAravallis, one of the oldest mountain ranges in the world, bisect Rajasthan diagonally. Guru Shikhar (1,722 meters) at Mount Abu is the highest peak.

b. Malwa Plateau: The Malwa Plateau is a significant volcanic plateau in central India that lies between the Aravallis and the Vindhyas.

c. Vindhya Range: The Vindhyas are a minor mountain range that runs parallel to the Narmada River south of the Malwa Plateau. They serve as a border between India's north and south.

d. Satpura Range: The Satpura Range is a rocky mountain range that spans Madhya Pradesh and Maharashtra to the south of the Vindhyas. This is the location of the Pachmarhi hill station.

C. Rivers and Drainage: The Peninsular area is home to older rivers that meander over wide valleys, including the Narmada, Tapti, Godavari, Krishna, and Kaveri. While other rivers flow eastward into the Bay of Bengal, the Narmada and Tapti rivers flow westward into the Arabian Sea.

D. Plateaus and Plains

a. Chota Nagpur Plateau: This plateau, which is part of Jharkhand and the neighboring states, is abundant in mica, iron ore, and coal.

b. Maharashtra Plateau: renowned for its rough topography, divided plateaus, and Deccan Trap-induced basaltic rock formations.

3. Extrapeninsular Geomorphic Features of India: The hilly areas in the north and northeast that resulted from tectonic activity related to the Indian Plate colliding with the Eurasian Plate are part of the extrapeninsular zone. Key characteristics include:

A. **Himalayas:** About 50 million years ago, the Indian and Eurasian plates collided to produce the Himalayas, a juvenile fold mountain range.

Three Major Ranges:

a. Greater Himalayas (Himadri): the tallest and northernmost range, which is home to peaks like K2 (8,611 meters) and Mount Everest (8,848 meters). At 8,586 meters, Kangchenjunga is the highest mountain in India.

b. Lesser Himalayas (Himachal Range): This range, which lies south of the Greater Himalayas, is home to well-known hill towns including Darjeeling, Mussoorie, and Shimla.

c. Shivalik Range: The foothills of the Himalayas are marked by the southernmost range, which is made up of loose sediments.

B. Indo-Gangetic Plain

a. Alluvial Plains: This area is one of the most productive and heavily inhabited in the world, having been formed by the deposition of sediments from the Ganges, Brahmaputra, and Indus rivers.

b. Extent: From the Brahmaputra in the east to the Indus in the west, the plain spans northern India.

c. Three Sections:

- Punjab Plain: It is formed by the Indus and its tributaries.
- Ganga Plain: It is formed by the Ganges and its tributaries.
- Brahmaputra Plain: It is formed by the Brahmaputra River.
- **Northeastern Hills (Purvanchal Range):**These hills, which comprise the Patkai, Naga, Manipur, and Mizo Hills, are found in northeastern India. These hills, which are a part of the outer Himalayas, are renowned for their heavy rainfall and thick forests.
- **Karakoram and Trans-Himalayan Ranges**

a. Karakoram Range: This range, which is north of the Himalayas, has peaks like K2 and the second-longest glacier in the world, the Siachen Glacier.

b. Ladakh and Zanskar Ranges: These mountains, which are a part of the Trans-Himalayan area, are distinguished by their rough topography, chilly deserts, and dry landscapes.

India's varied geomorphic characteristics are essential in determining the country's climate, biodiversity, and human settlement patterns.

Chapter 40
Physical Divisions of India and Their Characteristics

Introduction to the Physical Divisions of India

India's physical landscape is varied, ranging from vast plains and towering mountain ranges to dry deserts and verdant coastal areas. India's physical topography has a significant impact on its biodiversity, climate, natural resources, and human activity. This chapter examines India's main geographical divisions, their unique features, and their importance.

The Northern Mountains

A. The Himalayas

1. **Formation and Geography**
    a. **Formation:** Approximately 50 million years ago, the Indian and Eurasian plates collided to build the Himalayas.
    b. **Extent:** Stretch over 2,400 kilometers from Arunachal Pradesh in the east to Jammu & Kashmir in the west.
2. **Divisions of the Himalayas**
    a. **Trans-Himalayas:** It includes the Karakoram, Zanskar, and Ladakh ranges and is situated north of the Greater Himalayas.

b. **Greater Himalayas (Himadri):** The highest range, which includes peaks such as Nanda Devi, Kanchenjunga, and Mount Everest.

c. **Lesser Himalayas (Himachal):** Popular hill stations like Shimla and Darjeeling are located at a lower height.

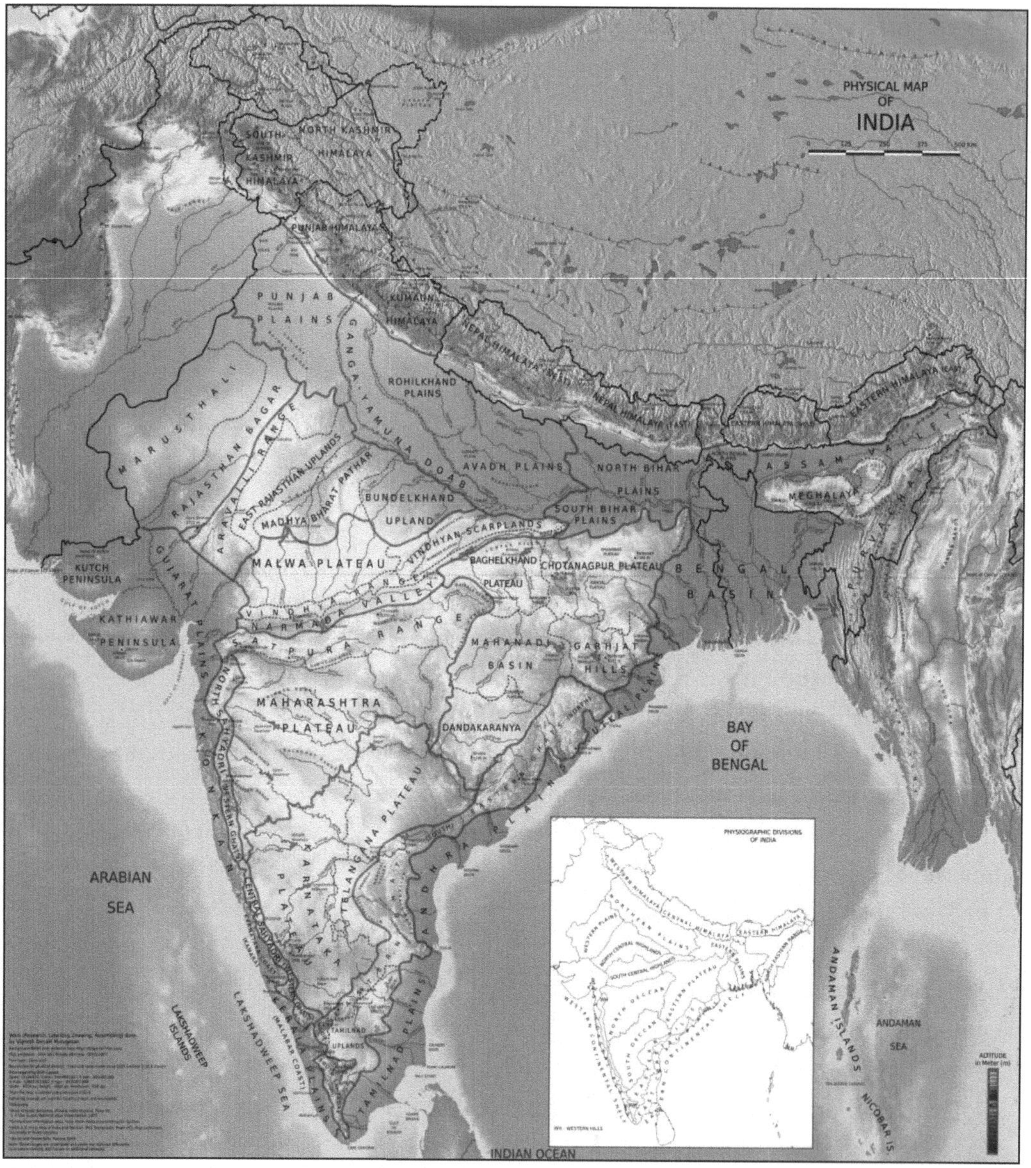

Figure 40.1: Physical Divisions of India(Reference: Wikipedia, *the Free Encyclopedia.*)

d. **Shivalik Hills:** The westernmost range is distinguished by a sequence of valleys known as "duns" and low altitudes.

3. **Characteristics**

a. **Climate:** Varies from subtropical in the lower sections to alpine in the upper ones.
b. **Vegetation:** Varies from alpine meadows to temperate woods.
c. **Significance:** Major rivers like the Ganges, Indus, and Brahmaputra originate here; it is rich in biodiversity and important for both tourist and cultural legacy.

The Eastern Hills

1. Geography and Extent

a. Location: Encompassing the Patkai, Naga, Manipur, and Mizo Hills in northeastern India.
b. Extent: Dispersed over the Meghalaya, Assam, Nagaland, Manipur, Mizoram, and Tripura states.

2. Characteristics

a. Climate: Due to the effect of the monsoon, it is often humid and experiences a lot of rainfall.
b. Vegetation: Dense woods that are home to a wide variety of tropical and subtropical species.
c. Significance: Vital to the production of tea, the preservation of biodiversity, and cultural variety.

The Northern Plains

A. The Indo-Gangetic Plains

1. Formation and Geography

a. Formation: Created by the Indus, Ganges, and Brahmaputra rivers and its tributaries depositing sediments.
b. Extent: Extend from the Assam valley in the east to the plains of Punjab in the west.

2. Divisions

a. Punjab Plains: Controlled by the Indus and its offshoots.
b. Ganges Plains: Stretching from West Bengal to Uttarakhand.
c. Brahmaputra Plains: Including portions of Arunachal Pradesh and Assam.

3. Characteristics

a. Climate: Mostly subtropical, having a monsoon season, cold winters, and scorching summers.
b. Soil: Alluvial soil that is fertile and perfect for farming.
c. Significance: One of the world's most densely inhabited areas; a significant agricultural region that produces sugarcane, rice, wheat, and other commodities.

The Peninsular Plateau

A. The Deccan Plateau

1. Geography and Extent

a. Location: Encompasses much of southern India and is bordered by the Satpura Range, the Eastern and Western Ghats.

b. **Extent:** Stretches from the Indian subcontinent's center to its southernmost point.

2. **Characteristics**

a. **Climate:** With distinct wet and dry seasons, it ranges from semi-arid to tropical.

b. **Vegetation:** Include grasslands, thorny scrublands, and tropical dry woods.

c. **Significance:** Rich in mica, coal, and iron ore, it is traversed by important rivers including the Kaveri, Krishna, and Godavari.

B. The Central Highlands

1. **Geography and Extent**

a. **Location:** Located north of the Plateau of the Deccan.

b. **Extent:** Comprises the Chotanagpur Plateau, Bundelkhand, and the Malwa Plateau.

2. **Characteristics**

a. **Climate:** Seasonal, subtropical climate.

b. **Vegetation:** A combination of meadows and deciduous woodlands.

c. **Significance:** Renowned for its abundance of minerals, including as bauxite, iron ore, and coal; it also produces wheat, legumes, and oilseeds.

The Indian Desert

A. The Thar Desert

1. **Geography and Extent**

a. **Location:** Situated mostly in the state of Rajasthan in northwest India.

b. **Extent:** Extends into Pakistan and has an area of over 200,000 square kilometers.

2. **Characteristics**

a. **Climate:** Arid, having high temperatures and little precipitation.

b. **Vegetation:** Sparse, made up of plants that can withstand drought (xerophytic).

c. **Significance:** Renowned for its distinctive desert vistas, rich cultural legacy, and commercial pursuits including mining and livestock husbandry.

The Coastal Plains

A. The Western Coastal Plains

1. **Geography and Extent**

a. **Location:** Stretch from Gujarat to Kerala along the Arabian Sea.

b. **Divisions:** The Malabar Coast (Kerala), the Kannad Coast (Karnataka), and the Konkan Coast (Maharashtra and Goa).

2. **Characteristics**

a. **Climate:** Tropical, having a lot of rain during the monsoon.

b. **Vegetation:** Mangroves, coconut orchards, and dense tropical woods.

c. **Significance:** Vital to tourism, agriculture (rice, coconut, spices), and fishing.

B. The Eastern Coastal Plains

1. **Geography and Extent**

a. **Location:** Stretch from West Bengal to Tamil Nadu along the Bay of Bengal.

b. **Divisions:** Stretch from West Bengal to Tamil Nadu along the Bay of Bengal.

2. **Characteristics**

a. **Climate:** Tropical, having cyclones and monsoons that occur seasonally.

b. **Vegetation:** Vegetation around the coast, such as casuarina and mangrove forests.

c. **Significance:** Major ports, the fishing industry, and agriculture are all supported by the fertile deltas of the Ganges, Godavari, Krishna, and Kaveri.

The Islands

A. The Andaman and Nicobar Islands

1. **Geography and Extent**

a. **Location:** Located on the Bay of Bengal, which stretches from Sumatra to Myanmar in the south.

b. **Divisions:** The Nicobar Islands in the south and the Andaman Islands in the north are the two major groupings.

2. **Characteristics**

a. **Climate:** Tropical, with high levels of humidity and precipitation.

b. **Vegetation:** Abundant biodiversity in dense tropical rainforests.

c. **Significance:** Numerous habitats, a prime position, and substantial tourist potential.

B. The Lakshadweep Islands

1. **Geography and Extent**

a. **Location:** Located off the southwest coast of India in the Arabian Sea.

b. **Extent:** Comprises 36 islands, with 10 inhabited.

2. **Characteristics**

a. **Climate:** Tropical, with humid weather and modest rainfall.

b. **Vegetation:** Sandy beaches, coral reefs, and coconut plantations.

c. **Significance:** Renowned for its tourism, marine biodiversity, and coral reefs.

Summary and Significance of Physical Divisions

India's geographical divisions—from the high Himalayas and the rich Indo-Gangetic Plains to the dry Thar Desert and verdant coastal regions—all have a distinct impact on the nation's topography, climate, wildlife, and human activity. Comprehending these distinctions is crucial for disaster relief, sustainable development, environmental preservation, and valuing India's abundant natural legacy. India is a country of unmatched variety and richness because of its diverse landforms, climates, and extensive agricultural, industrial, and cultural traditions.

References

1. Wikipedia contributors. (2024). Stylolite. In Wikipedia, The Free Encyclopedia. Retrieved from https://en.wikipedia.org/wiki/Stylolite.
2. Wikipedia Contributors. "Types of volcanic eruptions." Wikipedia, The Free Encyclopedia, Wikimedia Foundation, [Date of Last Revision]. Available at: https://en.wikipedia.org/wiki/Types_of_volcanic_eruptions.
3. Wikipedia contributors. (n.d.). Mare Orientale. In Wikipedia, The Free Encyclopedia. Retrieved [2023, August 19], from https://en.wikipedia.org/wiki/Mare_Orientale
4. Wikipedia contributors. (2023, August 19). Patterned ground. Wikipedia. https://en.wikipedia.org/wiki/Patterned_ground
5. Encyclopaedia Britannica. (n.d.). Pingo. In Britannica.com. Retrieved September 4, 2024, from https://www.britannica.com/science/pingo
6. Wikipedia contributors. (2023, August 24). Dune. In Wikipedia, The Free Encyclopedia. Retrieved August 27, 2024, from https://en.wikipedia.org/wiki/Dune
7. Encyclopædia Britannica. (n.d.). Ocean basin. Encyclopædia Britannica. Retrieved August 21, 2024, from https://www.britannica.com/science/ocean-basin
8. Wikipedia contributors. (2024, August 13). List of tectonic plates. In *Wikipedia, The Free Encyclopedia*. Retrieved from https://en.wikipedia.org/wiki/List_of_tectonic_plates
9. NASA Earth Observatory. "GRACE: Data Processing and Products." Earth Observatory, NASA, https://earthobservatory.nasa.gov/features/GRACE/page3.php. Accessed 27 Aug. 2024.
10. National Geographic Society. "Continental Drift." National Geographic, https://education.nationalgeographic.org/resource/continental-drift/. Accessed 27 Aug. 2024.
11. Wikipedia contributors. (2024, August 27). Plate tectonics. In Wikipedia, The Free Encyclopedia. Retrieved from https://en.wikipedia.org/wiki/Plate_tectonics
12. Wikipedia contributors. (2023, August 12). Hypsometry. In Wikipedia, The Free Encyclopedia. Retrieved August 27, 2024, from https://en.wikipedia.org/wiki/Hypsometry
13. Lincoln Learning Solutions. (2022, September 01) Formation of the Planets YouTube. https://www.youtube.com/watch?v=sCkhEu3lYNc&t=5s